Excel

YEAR 10

Science Revision Workbook

ESSENTIAL skills

Get the Results You Want!

PASCAL PRESS

Donna Bennett

Reprinted 2002, 2003, 2005, 2007, 2008, 2009, 2012
Updated in 2015 for the Australian Curriculum
Reprinted 2019, 2022

ISBN 978 1 74020 081 3

Pascal Press
PO Box 250
Glebe NSW 2037
(02) 8585 4050
www.pascalpress.com.au

Publisher: Vivienne Joannou
Project editor: Mark Dixon
Edited by Shelley Barons
Australian Curriculum updates edited by Geoffrey Thickett
Typeset by Scriptorium Desktop Publishing Pty Ltd and Kim Webber
Cover and page design by Dizign Pty Ltd
Printed by Vivar Printing/Green Giant Press

Contents

Introduction

This book is suitable for students studying the Year 10 Australian Curriculum Science course. It is particularly valuable for those students who are interested and motivated to seek an excellent result in Science.

This is a user-friendly revision workbook that students may use throughout the year to support classwork and help them with their revision for tests and exams.

The features of this book are as follows:

- The eight topics conform to the Australian Curriculum Science syllabus.
- Each topic consists of a concise summary of the key concepts in the topic.
- Each topic summary is followed by several pages of skill and knowledge practice questions to further enhance understanding.
- There are two sample Test Papers which bridge four topics and are aimed at the consolidation of knowledge.
- Answers to all questions are provided together with a complete glossary.
- Questions provide valuable revision before a test or exam.
- The book includes questions to interest the curious, capable student.

CHAPTER 1

EVOLUTION

Where did we come from?

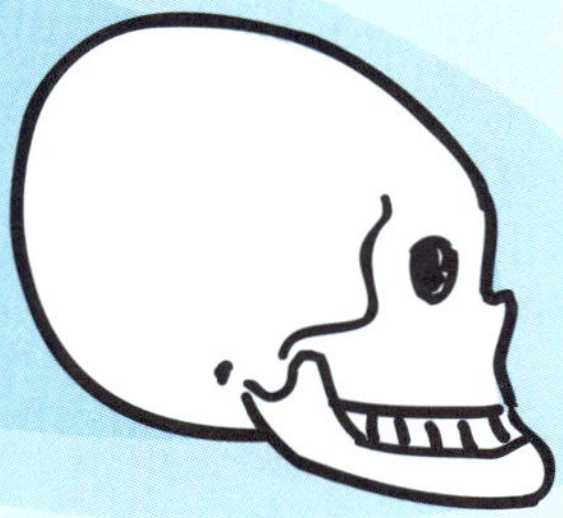

Variation

Variation is the natural differences between living things. Animals and plants that are produced by sexual reproduction always vary from their relatives. This is because each one has its own unique collection of genes. Variation is important because it enables a species to change slowly through the process of **evolution**.

A population of living things has three sources of variation:

- Immigration of individuals with new genes into a population.
- Recombination of genes during meiosis.
- Mutation.

Natural selection

Natural selection depends on evolution to create new forms of life. Any heritable characteristic that makes living things better suited to their way of life – their ability to survive and reproduce, relative to other members of the same **species** – will increase their numbers. The **adaptation** of organisms to the world around them affects not only their physical features but also behaviour and internal processes.

Humans selectively breed individuals with desired characteristics, for example domestic pets, crops and farm animals. Such artificial selection has created vast changes in some species over short periods of time.

The theory of evolution

Evolution is the process by which living things change as one generation succeeds another. Evolution is extremely slow, so many generations have to pass before any change is noticeable. The existence of evolution and the mechanism behind it were revealed by one of the world's great naturalists – Charles Darwin. From 1831 to 1836, Darwin sailed around the world on a survey ship called the HMS *Beagle*, describing hundreds of plants and animals on the way.

Darwin published his theory of evolution in his famous book *On The Origin of Species*.

Darwin's theory of evolution is based on three observations:

- That all living things vary.
- That all living things can pass on their characteristics.
- That all living things are involved in a struggle for survival.

In this struggle some organisms do better than others, and so survive to produce more offspring. Therefore their characteristics become more common in later generations.

This natural selection occurs over many generations and results in change, or evolution.

Evolution is happening all the time but it is not easy to see as changes build up slowly. Natural selection gradually alters the complete collection of genes, or the **gene pool** present in the species.

In recent years more and more evidence has been found to support Darwin's theories. In particular, scientists have found that the DNA of all living things is basically the same, and it is translated into proteins by the same genetic code. This strongly implies that all living things have evolved from a common ancestor.

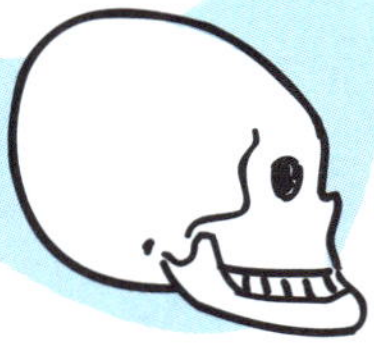

The struggle for survival

All living things have the potential to produce vast numbers of offspring. However, their numbers usually stay the same because their offspring depend on limited resources, and have to struggle, or compete, to survive. This competition allows natural selection.

Extinction

During evolution, some alleles become more common and others are lost forever. Some species grow in numbers and others decline, eventually becoming endangered.

When a species dies out all together it is said to become **extinct**. Fossil records show that many species have become extinct. Humans have caused an enormous impact on species' gene pools through hunting and, more recently, habitat destruction.

Fossils

Much of our knowledge about the history of the Earth comes from fossils. Fossils are the preserved remains, impressions or traces of past life on Earth. Organisms that died in or near water had a better chance of fossilisation.

The main types of fossils are:

- Fossils that have been replaced by rock. These are usually made of the hard parts of organisms like shells, bones, teeth and wood. This is called **petrification**.
- Fossils formed when minerals from the rocks replace the organism's body materials. This is called **mineralisation**.
- Fossils of soft organisms, like jellyfish and plants, mostly forming **moulds**. As the original material decays it leaves an impression in the rock.
- **Footprints** that have been preserved in mud – these are called castes.
- Actual remains, like **bones**. Some animals have been reconstructed using their bones.

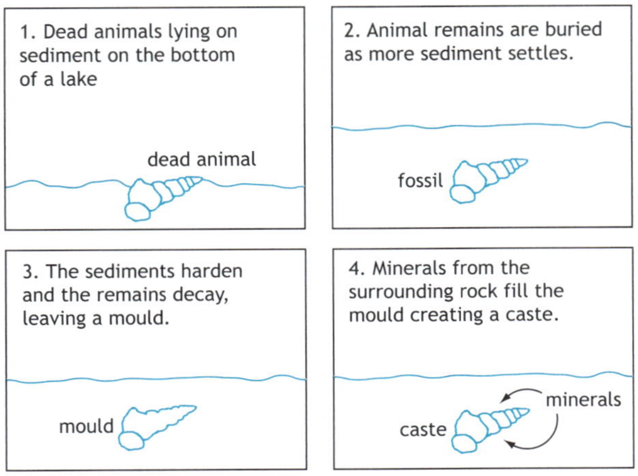

The formation of a mould

This evidence is called a **fossil record** and scientists who study fossils are called **palaeontologists**.

Radioactive dating can determine the age of rocks. Knowing the age of the rock helps scientists to determine the age of any fossils embedded in it. The age of fossils can also be estimated by using the half-life of **carbon-14**.

Some fossils lived only at a particular time in history and can be used to tell the age of the rock in which it was fossilised. For example the Trilobite lived in the Palaeozoic Era and then became extinct.

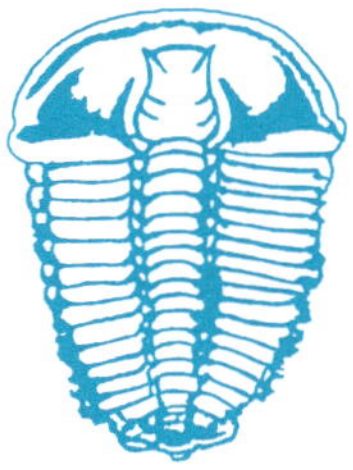

Trilobite

The history of life

The Earth has existed for about 4.5 billion years and the first living things appeared 3.8 billion years ago. In that time many life forms have flourished and disappeared, leaving fossils as evidence.

Geological time scale

Geologists divide the Earth's history into three different time spans: era (longest), periods and epochs (shortest).

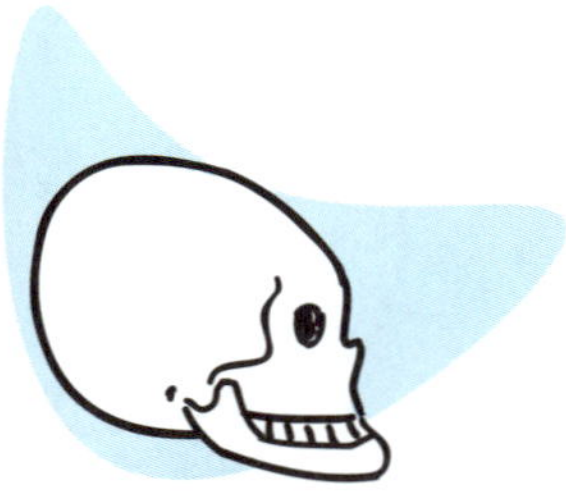

Each interval marks a particular stage in our planet's history.

Record of past life (M = millions of years)

Years (M)	Era	Period	Animals
2	Cainozoic	Quaternary	Homo sapiens
70		Tertiary	Primates
135	Mesozoic	Cretaceous	Flowering plants
180		Jurassic	Dinosaurs First bird
225		Triassic	Dinosaurs Mammals
270	Palaeozoic	Permian	Ferns and conifers
300		Late carboniferous	Reptiles
350		Early carboniferous	Vast forests
400		Devonian	Insects Amphibians Vascular plants
440		Silurian	Land plants Fish with jaws Coral
500		Ordovician	Molluscs Crustaceans
600		Cambrian	Invertebrates Trilobites
		Precambrian	Simple plants and animals

Ice ages

In the past the Earth's climate has at times become so cold that ice from the poles and mountains has spread over large areas of land. This is called an ice age and can last for over a million years. Ice ages can dramatically affect an evolutionary path.

Mass extinction

During the Earth's history large numbers of species have become extinct in short spaces of time. Scientists believe there have been five mass extinctions so far. The greatest mass extinction ever known was between the Permian and Triassic periods, 245 million years ago. About 65 million yeas ago the dinosaurs died out in the Cretaceous period. No-one knows what caused these mass extinctions, however, some scientists believe there was a massive climate change at the end of the Cretaceous period caused by the impact of a meteorite.

Human evolution

Humans have existed for over a million years and, like all living creatures, humans have been shaped by evolution. Before we evolved, several human-like species existed. Scientists are still trying to piece together where and how they lived and which ones were our ancestors.

Humans belong to a family of primates called **hominids**. Unlike other primates, like apes, hominids have large brains and walk upright on two legs. Within the hominid family, humans form part of a smaller group, or **genus**, called Homo. The genus Homo also includes our extinct ancestors. The first hominids appeared on Earth about 3 – 4 million years ago.

Walking on two legs (**bipedal locomotion**) freed up their hands, which became especially adapted for gripping things. Eventually they were able to make tools.

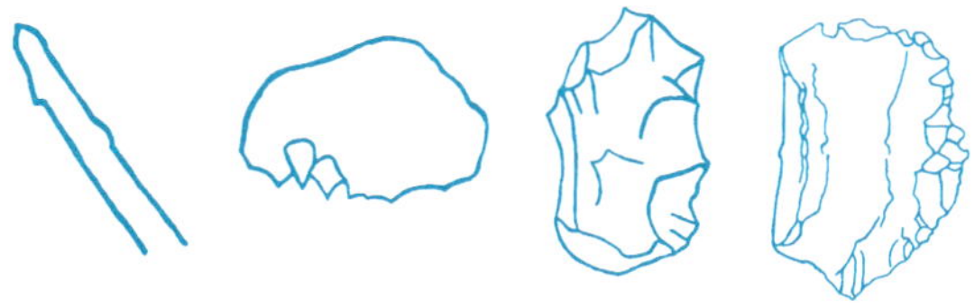

Tools of early hominids

Fossil evidence shows that hominids have existed for over 15 million years.

- *Australopithecines* is a genus of early hominid known from fossils found in Africa. This is the oldest hominid known and may have walked in Africa 4 million years ago. They walked on two legs and had small brains. 'Lucy' is the fossilised skeleton of a female *australopithecines* who died over 3 million years ago.
- *Homo* is the genus which includes the human species and ancestors of humans. *Homo* appeared about 2 million years ago in Africa and Asia and had large brains. All except humans have become extinct.
- *Neanderthal man* is a form of human that died out about 30 000 years ago. *Neanderthals* lived in Africa, Europe and Asia. They looked like modern humans, had large brains and used tools.

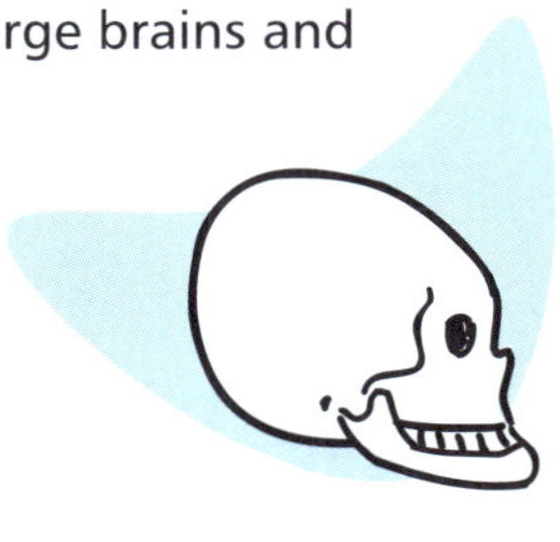

- *Homo sapiens* is the current form of the human species, which appeared about 300 000 years ago. Originally there were two forms of *Homo sapiens*, but Neanderthal man died out.

Diversity

The fossil record shows that life underwent a massive explosion of diversity about 500 million years ago in the Cambrian era. All the major body plans that represent life on the planet today were invented in a frenzy of evolutionary innovation, including amphioxis, the founder of phylum chordata. The reason for this explosion has not been explained but some scientists believe there was a major increase in oxygen levels around this time.

Although living things first appeared long before, they were small, soft-bodied organisms and left few traces.

The biological diversity on the planet today is close to the peak in Earth's history. We are one species among countless millions – the richest array of life the Earth has ever hosted.

Glossary

Adaptation – A process that makes living things better suited to their way of life.

Bipedal locomotion – Walking on two legs.

Evolution – A gradual genetic change in a group of living things.

Extinction – The permanent disappearance of a species.

Fossil – The remains of a living thing that have been preserved.

Gene pool – All the genes in a group or species.

Hominid – A family of primates that includes humans and the immediate ancestors of humans.

Natural selection – The process by which many different natural factors, from climate to food supply, steer the course of evolution. Living things better suited to their environment have more offspring.

Species – A group of living things that can breed together in the wild.

Variation – The natural differences between living things.

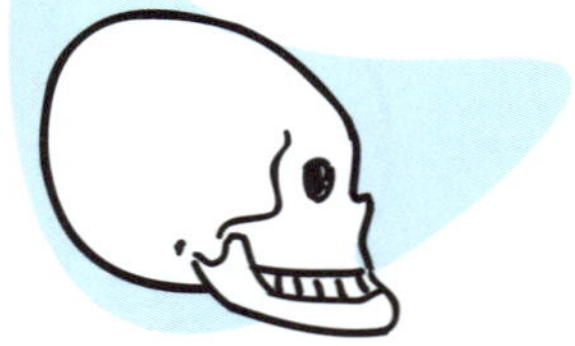

Science skills

Question 1

The following table displays the presence and abundance of different vertebrates during the various time periods.

Millions of years ago	Era	Period	Vertebrates
2	CAINOZOIC	Quaternary	
70		Tertiary	
135	MESOZOIC	Cretaceous	
180		Jurassic	
225		Triassic	
270	PALAEOZOIC	Permian	
300		Late Carboniferous	
350		Early Carboniferous	
400		Devonian	
440		Silurian	
500		Ordovician	
600		Cambrian	
		Precambrian	

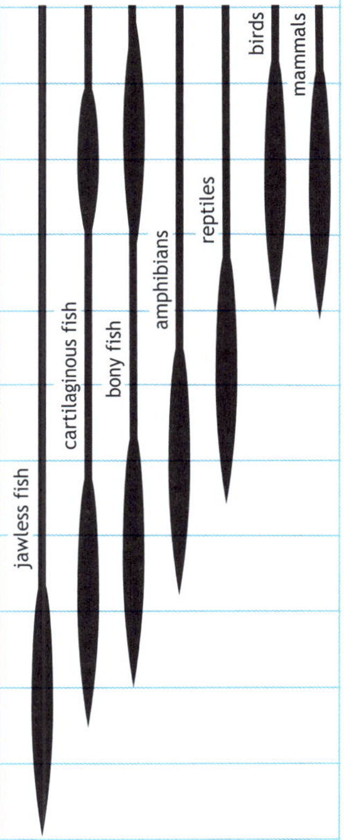

a In which period did mammals emerge?

b Which periods had the greatest numbers of amphibians?

c Which animals are at their maximum number now?

d The populations of which two groups varied considerably over the past 600 million years?

e Which animals emerged in the Ordovician period and lasted in large numbers only until the Early Carboniferous period?

f How many years ago was the Jurassic period?

g In which era was the Jurassic period?

h Which animals dominated the Earth in the Jurassic period?

Question 2

The following structures are the limbs of several animals. All the limbs have five parts. What part of Darwin's theory does this information support?

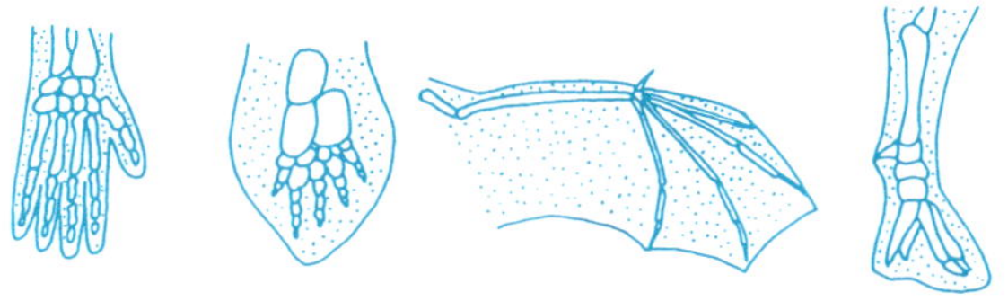

Question 3

The information over the page shows the geological time range of certain organisms.

a Fossils of these organisms are all found in one particular sedimentary layer. During which period was this sedimentary layer deposited?

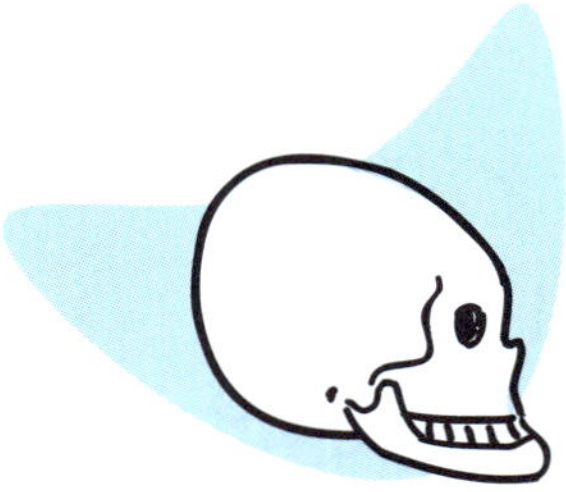

b Which fossils would you expect to find in a rock from the Ordovician period?

Geological periods

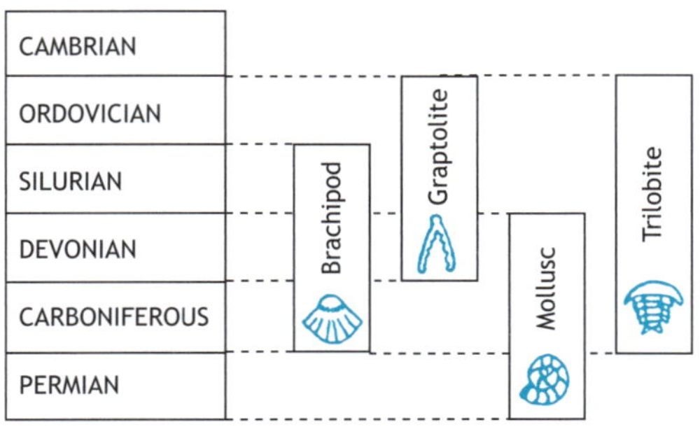

Question 4

Put the following diagrams into the correct order to explain the formation of a fossilised skeleton.

(a) Wind, rain and frost slowly wear the rock layers away.

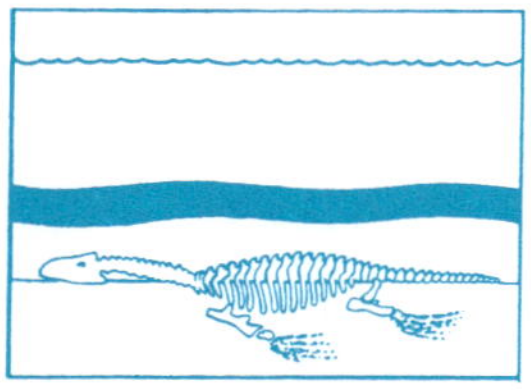
(b) Now only the skeleton of the animal is left buried in the mud.

(c) Fossilised skeleton begins to decay on the rock face.

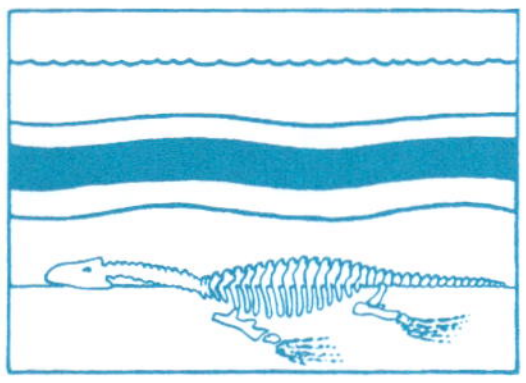
(d) The heavy layers of mud press down and harden into rock.

(e) The plesiosaur dies in a lake, and its body sinks to the bottom.

(f) Movements of the rock bring the skeleton back towards the surface.

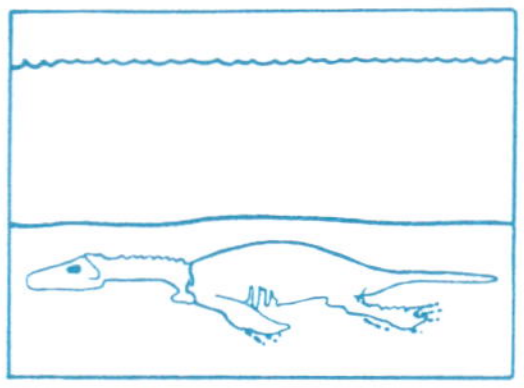
(g) The body rots, and is slowly covered with layers of mud.

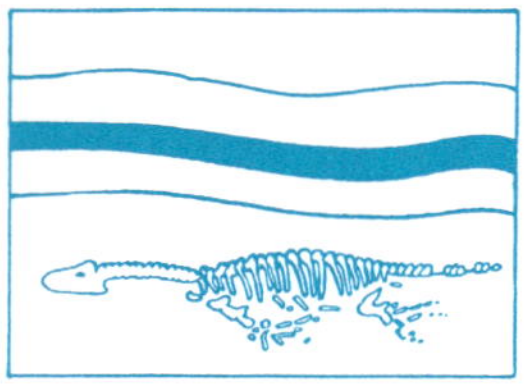
(h) The fossilised skeleton is left in the rock layers.

Question 5

The following graph shows that significant extinctions occurred over the last 26 million years. The five major ones are marked.

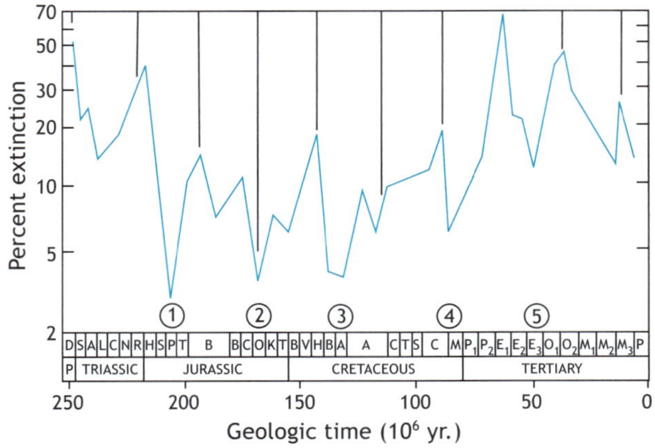

a Which extinction was the one that destroyed the dinosaurs?

b Which extinction caused the greatest percentage loss of organisms?

Question 6

The diagram over the page represents the number of living species of all kinds of organisms currently known.

a What is the total number of species represented?

b Which two groups dominate the diversity of the living organisms to date?

c Where do most of the species live – land, air or sea?

d How many species of protozoa have been discovered?

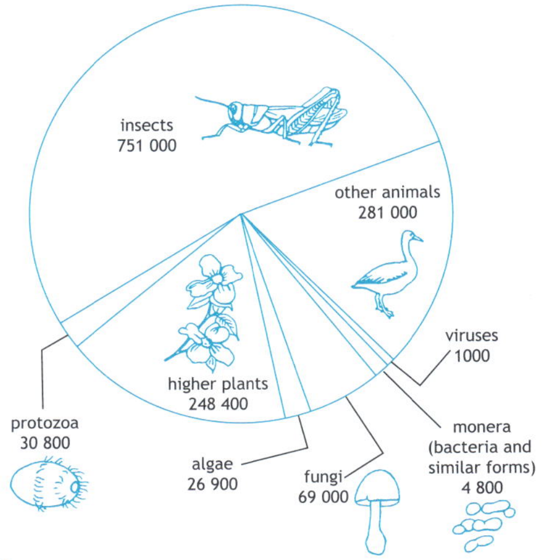

Question 7

Many lands suffered loss of large animal species during the Pleistocene period, but at different times. Losses in certain countries coincided with the colonisation of humans. Study the following graphs and answer the questions.

a When was North America colonised?

b What percentage of large animals has Africa lost?

c What percentage of large animals was lost after the colonisation of Australia?

d Which country has been colonised the longest?

e Which country was colonised most recently?

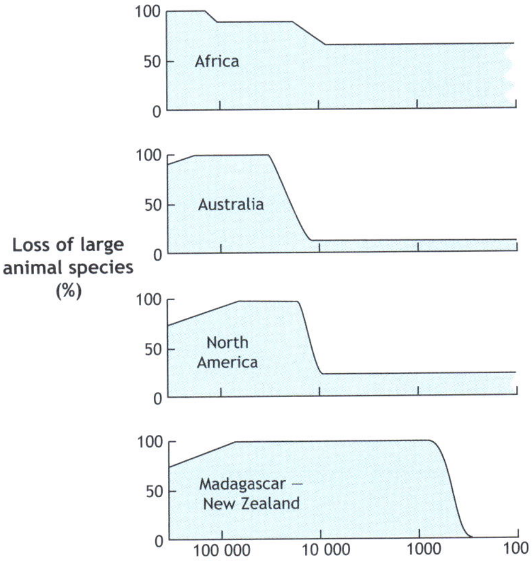

Question 8

Study the fossil record in the drawings of road cuttings below.

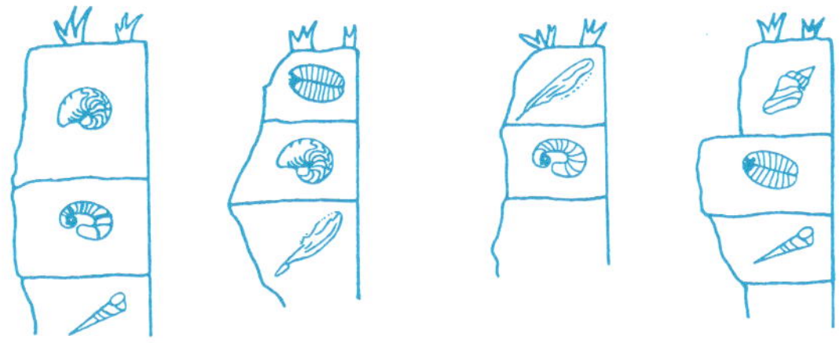

a b c d

There has been earth movement since these rock layers were formed, but in some areas erosion has occurred.

a Which road cutting contains the most recent fossil?

b Which layers are the oldest?

Question 9

Match the following fossils with the correct description.

a Very long tail, long arms and small hips.

b Tall spines from the backbone.

c Very large skull, short neck and a thick tail.

d Small head, very long neck and thick legs.

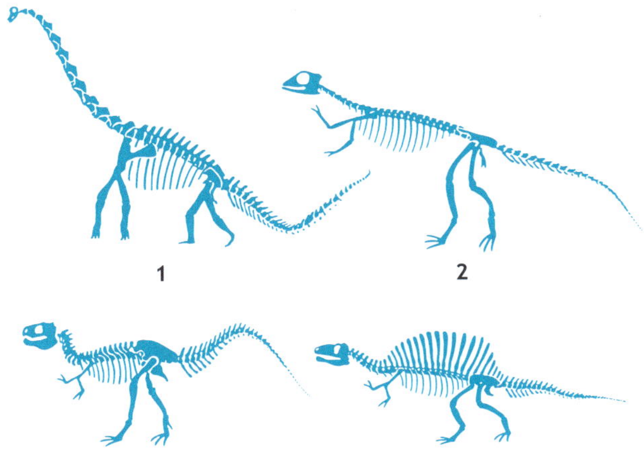

Question 10

Look carefully at the diagram below, which shows the particular time in Earth's history that various invertebrates existed.

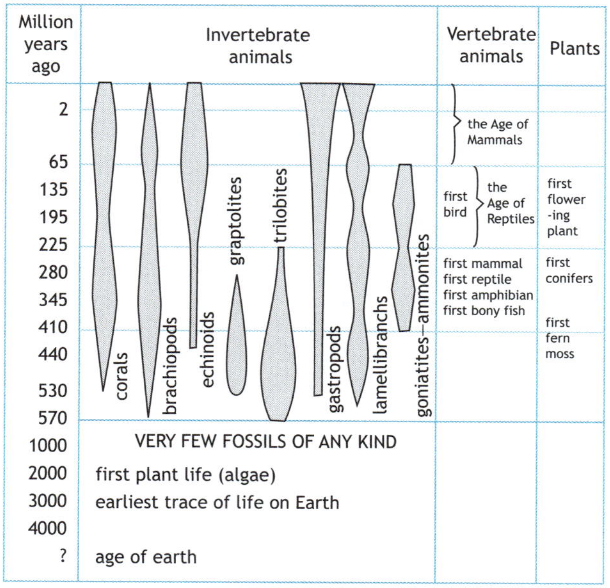

a Which invertebrates are now extinct?

---

b If you found a rock with a trilobite fossil in it, what could you say about the rock's age?

---

c What are the oldest rocks that could contain coral fossils?

---

d When did gastropods begin to become increasingly common?

---

e If a museum artist was asked to draw the plants that dominated the Earth at the time when the first dinosaurs, mammals, bony fish and amphibians were appearing, what would he draw?

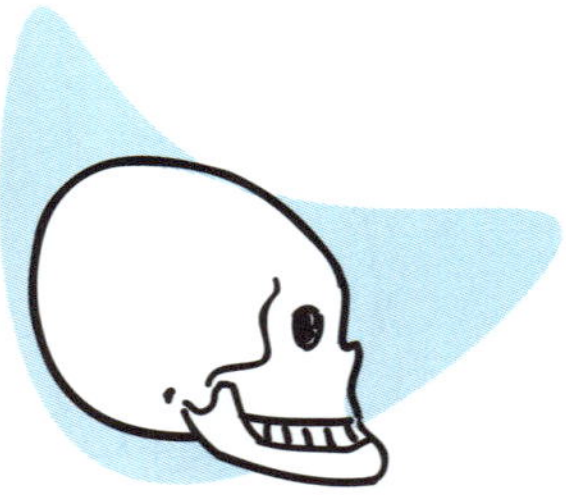

Question 11

The following fragment of a fossilised skull was found.

Which of the sample skulls does it most closely resemble?

A

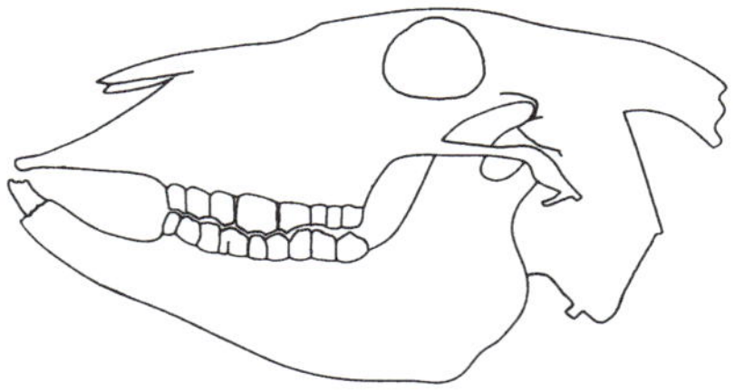

B

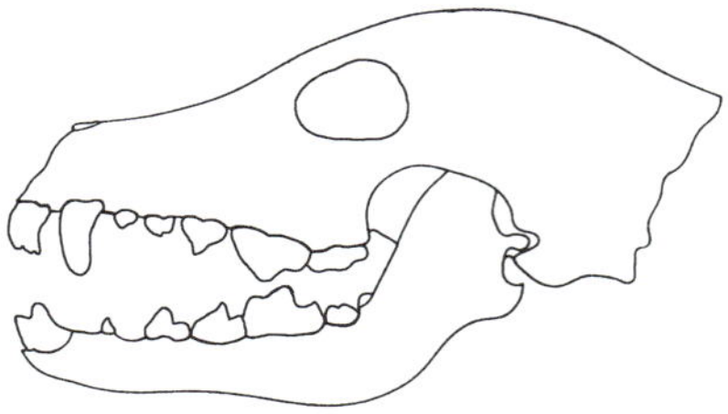

Sample skulls

Question 12

Put the following human ancestors in the correct chronological order.

Lucy

Neanderthal man

Homo

Australopithecines

Homo sapiens

Question 13

As environments change, so do the selective factors that affect the survival of organisms. Those organisms best suited to survive and breed in the changed environment pass their genes on to the next generation.

In Australia, wild rabbits were controlled using the myxoma virus. This virus kept the rabbit population under control for while.

Study the following graph and answer the questions.

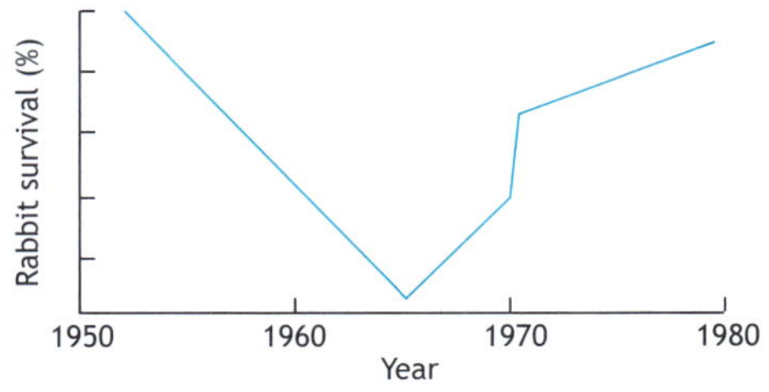

The Genetic Resistance of Wild Rabbits

a What happened to the rabbits in the 1950s and 1960s?

____

____

b Based on the information above, why do you think this happened?

____

____

c What process has been operating since 1965?

____

d Why was the rabbit population almost the same in 1980 as it was in 1950?

____

____

Question 14

The fossil record shows that the ancestors of the horse lived 60 million years ago. They were very small and had four toes on each front foot.

Scientists believe that the horses adapted to the need to run faster by changing the structure of the front leg.

a Put these diagrams into the correct order.

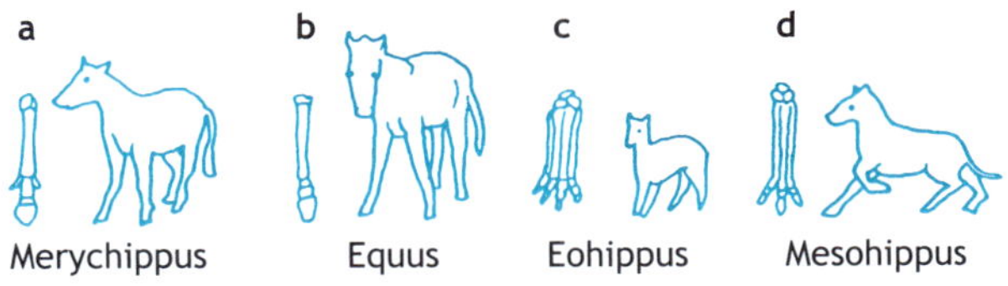

b How many toes on each foot does modern horse have?

____

c What is the name for this process of adaptation?

____

Revision questions

Question 1

If dinosaurs trod in the mud at the shore of a lake and the footprint hardened in the sun, it would become a ________ if it filled with sand or mud, which hardened over time.

Question 2

The fossil record provides evidence of evolution.

a What are two disadvantages of using the fossil record as a tool for studying evolution?

____

____

____

____

b What are two advantages of using the fossil record as a tool.

____

____

____

____

Question 3

If we know when a certain organism lived, it is possible to tell the age of a rock in which a fossil of that organism was formed.

Fill in the missing words.

A typical ________ fossil is the trilobite. Trilobites lived in the oceans in the ________. At the end of this era they became ________. If a rock contains a trilobite we can conclude that the rock was formed during this era.

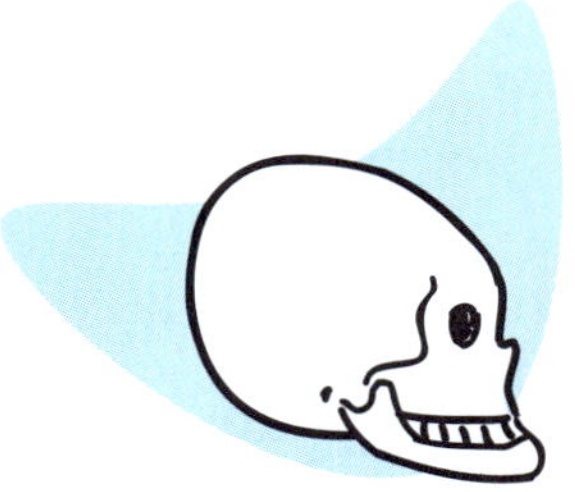

Question 4

a During which era did flowering plants become common on Earth?

____

b In which era were there vast forests of conifers?

____

c In which period did ferns appear?

____

d In which period did vascular plants appear?

____

e In which period did the very first plants appear?

____

Question 5

Fill in the correct words to complete the following information.

Dinosaurs were __________ that appeared about ______ million years ago and became _________ about ____ million years ago. Scientists are not sure why dinosaurs became extinct. One theory suggests that an __________ impact caused vast changes in __________ and dinosaurs were not able to __________ quickly enough.

Question 6

a Which family of primates do humans belong to?

____

b What are the main differences between humans and other primates?

____

____

c What advantage did walking on two legs give to humans?

____

____

d How long do scientists believe that hominids have existed on Earth?

____

Question 7

Complete the following information by filling in the correct words.

________ __________ was an English naturalist living in the 1800s. As he travelled the world in the ship The _________, he described enormous numbers of ________ and _________. These experiences helped him form the theory of __________. This theory states that all present living creatures gradually developed from __________ ancestors. He published his theory in a famous book called *On the* ____________ *of* ____________.

Question 8

a What happens to a species if some alleles are lost?

____

b What do we call it if a species dies out altogether?

____

c What is an endangered species?

____

____

d What is the main cause of species loss today?

____

Question 9

a To which group of vertebrates do the following extinct animals belong to?

____

b State possible reasons why these two animals became extinct?

____

c What is the name of the ancestor of vertebrates?

____

Woolly mammoth

Brachiosaurus

Question 10

What type of fossil is formed by the process illustrated below?

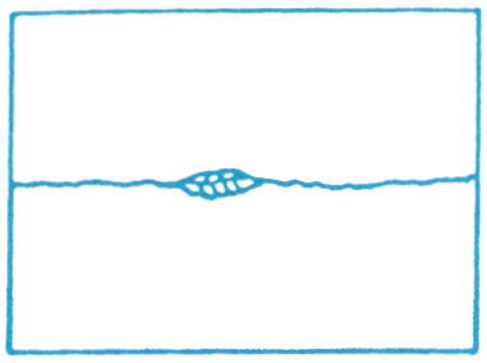

Plant leaf lying on sediment.

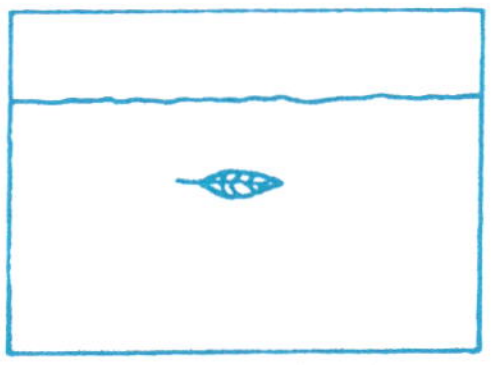

Leaf is buried under more sediment.

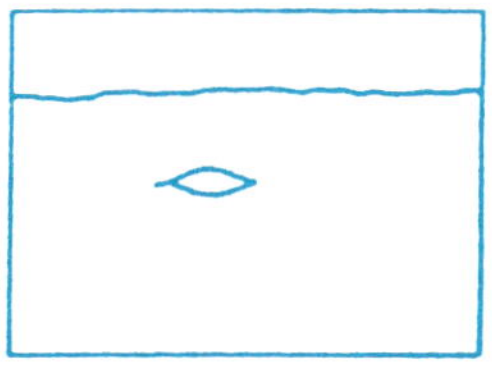

Sediment hardens, forming rock. The leaf decays and leaves a hole in the rock.

Question 11

a What is meant by the word 'evolution'?

b Darwin's theory of evolution was based in part on the term 'adaptation'. What does this term mean?

c Charles Darwin believed 'That evolution by natural selection was the result of the survival of the fittest individuals in a population'. What did Darwin mean by this statement?

Question 12

a What process forms a fossil in this way?

A fish lives in an ancient sea.

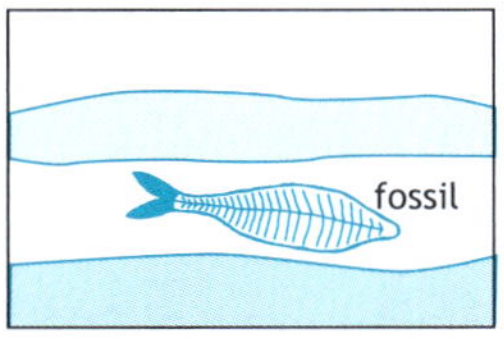

It dies and is covered by sediment. Gradually, the sediment form rock.

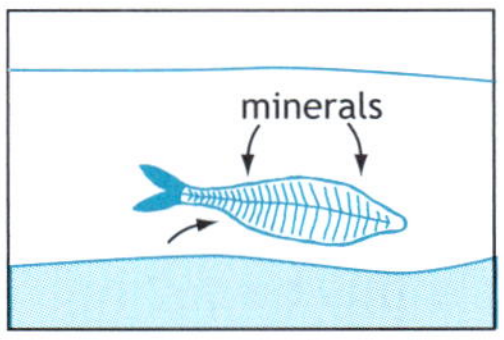

Minerals from the rocks replace the fish's body materials.

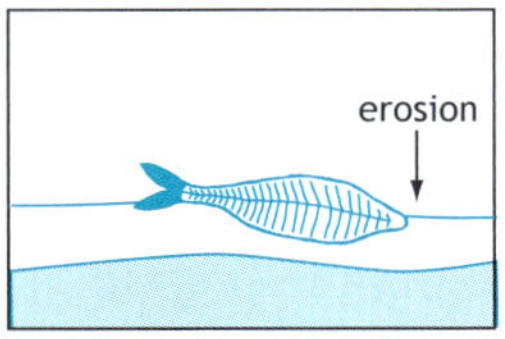

The rocks are eroded, leaving the fossil exposed.

b What it the process that involves replacement of hard parts of an organism by rock?

Question 13

a What is the name given to the major global events involving the spread of ice from the poles over large areas of land?

b What slow natural process do these events interrupt?

Question 14

The biological diversity of the Earth today is close to the peak in Earth's history.

a How many years ago was there an 'explosion' of diversity?

b In which era did this occur?

c What is one theory used to explain this explosion?

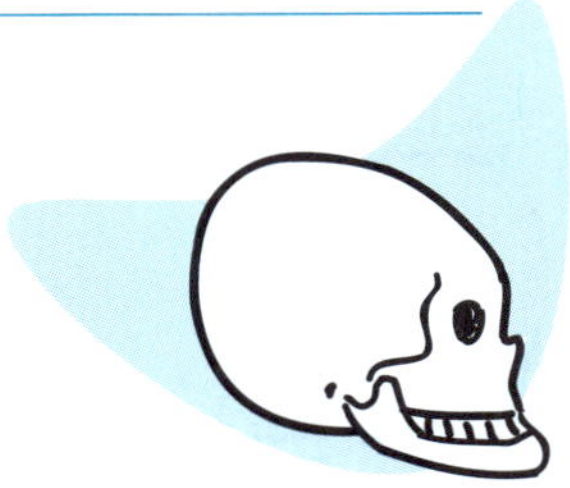

Question 15

a What are the three different time spans used to describe our geological history?

____

b Which one is the longest?

____

c Give an example of an era and a period.

____

____

Question 16

a Fossils are more common in Tertiary rocks than Silurian rocks. What might this be so?

____

____

____

b Marine fossils are found in abundance around Mt Isa. How could this have happened?

____

____

____

c In what order do the following animals appear in the fossil record?

mammals, amphibians, dinosaurs, fishes

____

Question 17

Fossils are most likely to be found:

a in the sand of deserts.

b in rocks that were formed by the cooling of lava from volcanoes.

c in rocks formed from the mud and sand near lakes and seas.

Question 18

Complete the following information by adding the correct words.

The modern technique of __________ ________, allows the _____ of rocks to be determined. This technique uses _________ rates of naturally occurring radioactive substances like __________. Knowing the age of rocks enables scientists to work out the age of a ________ embedded in the rock.

Question 19

a List the three possible sources of variation within a population of organisms.

____

____

____

b Why is variation within a species important?

____

____

____

Question 20

Number the following life developments in the record of past life, from the oldest event to the most recent.

- First reptiles appear, many large insects.
- Dinosaurs die out.
- Reptiles thrive, first mammals.
- Humans dominate, domestic animals.
- Early hominids.
- Mammals, modern, flowering plants.
- Many sea creatures.
- No life.
- Amphibians, ferns and conifers flourish.
- Giant dinosaurs, first birds, first flowering plants.

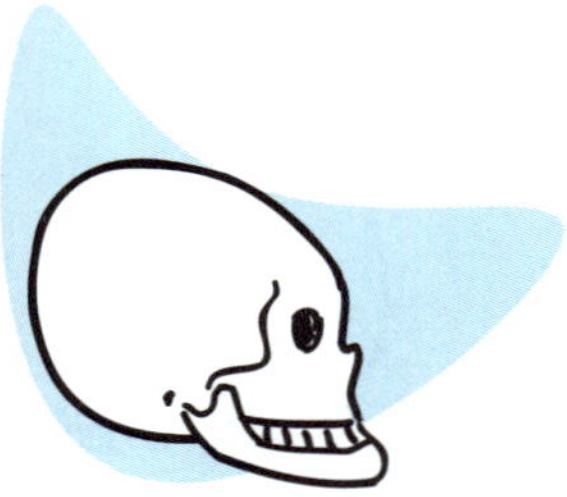

CHAPTER 2

GENETIC INHERITANCE

What part of your body contains more information than 100 computer disks?

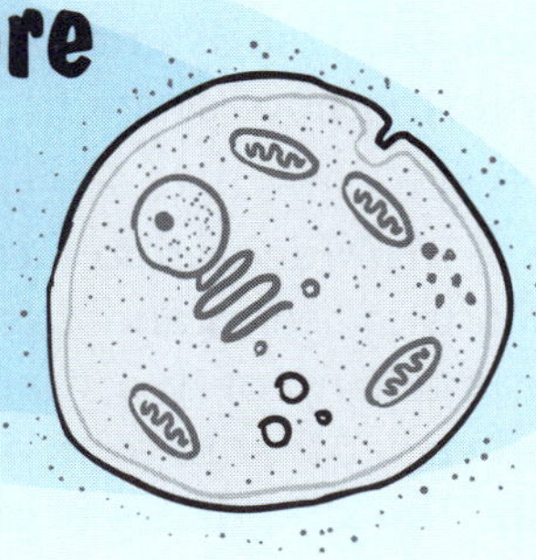

The flow of biological information

Chromosomes

Almost every cell in all living organisms has a set of **chromosomes**. These long thread-like structures store the DNA molecules that contain all the information needed to make cells work. Chromosomes pass on genetic material from one generation to another whenever a cell divides.

The genetic code

Body cells use the **genetic code** or **genome** to interpret the information stored inside **DNA**. A molecule of DNA contains a huge sequence of chemicals, called bases, arranged in a particular order.

The code consists of 64 possible sequences in which any of the four nitrogen-containing bases (adenine, thymine, guanine and cytosine) can be arranged. The genetic code tells a cell how to convert the DNA's sequence of bases, or each group of three bases, into a sequence of **amino acids** to make the proteins that build and control cells. Without the code, the genetic information inherited from parents would be meaningless.

Codon

The 'words' used in genetic code are called codons. Each **codon** is made up of three base pairs, arranged in any order, and has an exact meaning. A codon specifies a particular amino acid, or shows where a particular chain of amino acids should start or stop.

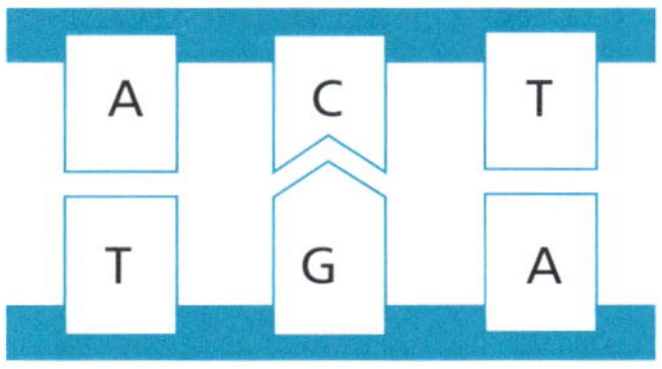

A = Adenine
U = Uracil
G = Guanine
C = Cytosine

A codon

Genes

The DNA in human cells stores between 50 000 and 100 000 genes. Each gene contains codons that instruct cells how to make certain proteins. By making different proteins at different times, genes can act as switches to alter the way cells work. Genes are the basic units of heredity and are passed on to the next generation without being divided into smaller units. However, each gene can exist in two different forms. These are called **alleles**. Every living thing produced by sexual reproduction has a unique set of alleles.

Nucleic acids

Nucleic acids are organic compounds made up of small molecules called nucleotides. There are two types of nucleic acid – DNA and RNA. **DNA** carries genetic information and is stored inside the nucleus. **RNA** acts as a 'shuttle service', copying the DNA's information and carrying it to where it is put into action.

A **nucleotide** is a molecule made up of a sugar, a base, and a phosphate group. The sugars link up to form a nucleic acid's 'backbone' and the bases make up its genetic information.

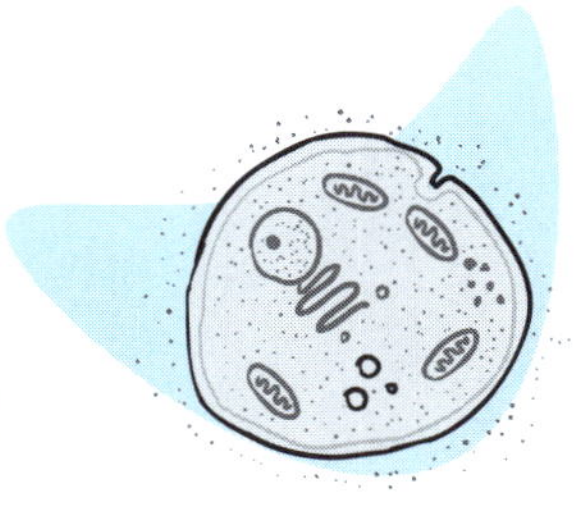

DNA

The structure of DNA was worked out in 1953 by James Watson and Francis Crick, in what became one of the most famous scientific discoveries of modern science.

A molecule of DNA (deoxyribonucleic acid), can contain millions of atoms and is found in the cell nucleus. It is made up of two strands, which twist around each other to form a shape called a **double helix**. The strands are held together by the bases. The exact sequence of these bases forms genetic information that varies from one living thing to another.

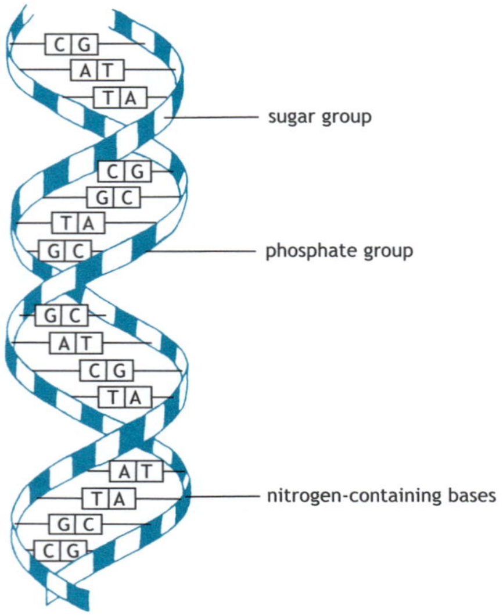

Double helix

Bases always pair up precisely. These are called **base pairs**. Adenine always pairs with thymine, and cytosine always pairs with guanine. This precise pairing of bases provides a complementary copy of the first helix strand.

RNA

Molecules of RNA (ribonucleic acid) are found in the nucleus and the cell cytoplasm. They are usually single threads and carry out several important tasks in making proteins. RNA contains the base uracil instead of thymine. Messenger RNA copies parts of the DNA's code and transfer RNA uses this information to link up the molecules that form the new proteins.

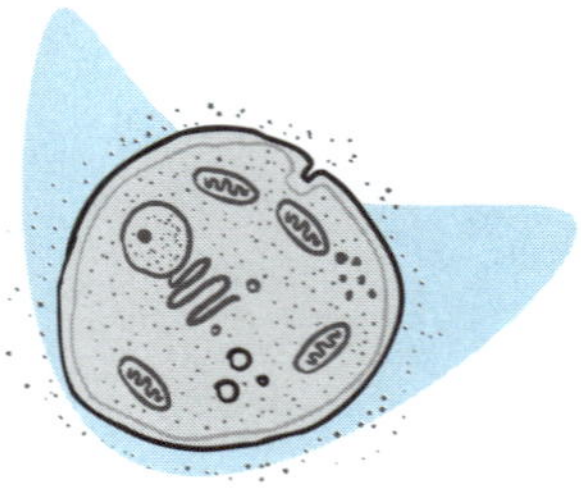

Replication

When a cell divides, its DNA molecules copy themselves, or replicate, in order to pass on a set of instructions to each new cell. During replication, each DNA molecule 'unzips' itself, so that its strands separate. The two strands then form complementary copies of themselves. Two new DNA molecules are produced, each with one old strand and one new strand.

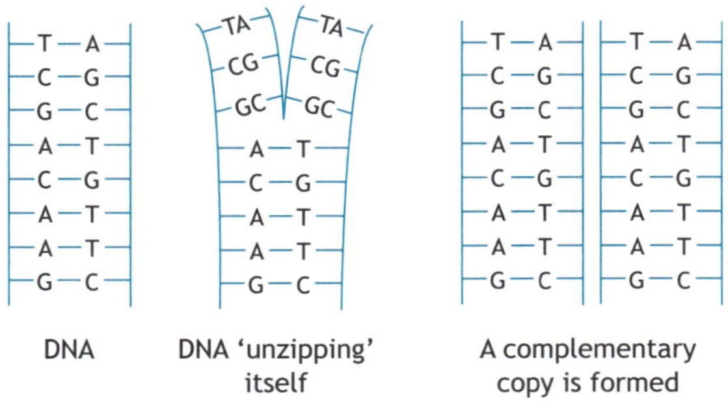

DNA replication

Making a protein

Cells can make proteins only by following the instructions of a gene. There are two stages in making a protein.

Stage 1 – Transcription

The two strands of DNA pull apart. The RNA makes a copy of the information on one of the strands that has separated. The DNA then closes up again.

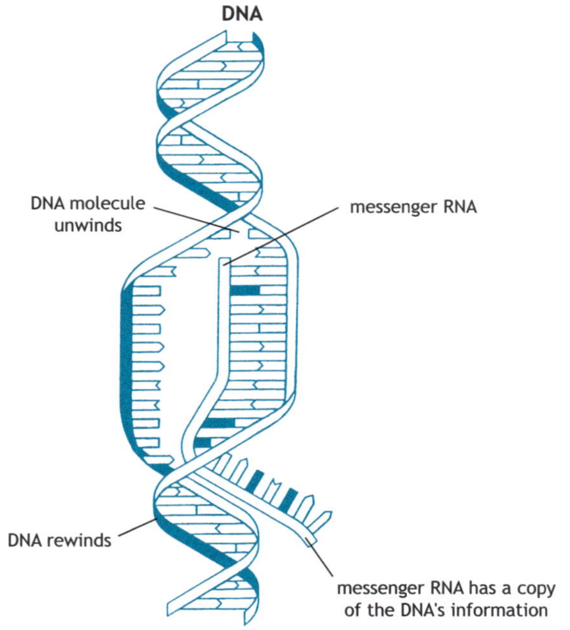

Transcription

Stage 2 – Translation

Chemical clusters called **ribosomes** now move along the new strand of messenger RNA, reading the strand's instructions one codon at a time. With the help of transfer RNA, the ribosomes convert the sequence of codons into the required sequence of amino acids, thus forming a new protein.

Heredity

This is the link between successive generations of living things. Living things produce offspring that are similar to, but not the same as, themselves by passing on genes for particular characteristics through sexual reproduction. Each individual has a unique selection of genes taken from both parents. This **variation** is important because it allows a species to change slowly through evolution.

Genotype

The **genotype** of a living organism is the genetic information it inherits from its parents through sexual reproduction. It includes genes or alleles that affect such factors as shape, size, colour and function. Some alleles are not expressed in every generation. The genotype interacts with the environment to produce the **phenotype** or visible characteristics of an organism.

Alleles

In cells there can be two sets of matching or homologous chromosomes, which contain pairs of genes called alleles. These are alternative versions of the same gene. For example eye colour could have two alleles, one for brown eyes and one for blue eyes. They are always in the same position, or **locus**, on the chromosomes.

- A **dominant allele** is the allele that will most likely express itself. For example if brown eyes are dominant over blue eyes, then this will be the allele that is expressed. Dominant characteristics are written as capital letters.
- A **recessive allele**, partnered by a dominant allele, is usually masked by the dominant one. However, it is still part of the genotype and can be passed on to the next generation. It will only be expressed if it is later partnered by an identical recessive allele. Recessive characteristics are notated with lower case letters.

Mendelian ratio

It has been shown that characteristics are inherited in a predictable and mathematical way. These patterns are called **Mendelian ratios** after the Austrian monk, Gregor Johann Mendel, who discovered them in 1866. They can be used to show how characteristics are passed from one generation to another.

For example, the inheritance of flower colour.

Two flowers, a red one and a white one, cross. The red is the dominant colour. How many of each colour will appear in the next generation?

red flower	white flower
RR	rr
(two dominant alleles)	(two recessive alleles)

First Generation

Rr	Rr	rR	rR
Red	Red	Red	Red

All the offspring in the first generation are red (same phenotype) with the same genotype.

Second Generation

red flower	red flower
Rr	rR

Offspring:	Rr	RR	rr	rR
	Red	Red	White	Red

Ratio is red (3) : white (1).

So three-quarters of the offspring in the second generation are red.

- **A homozygous cell** is a cell that has identical alleles that control its particular features. They are either both dominant or both recessive.
- A **heterozygous cell** is a cell that has one dominant and one recessive allele for its characteristics.
- During sexual reproduction genes are shuffled like cards in a pack. Alleles that are close to each other on the same chromosome often stay together. For example, in humans the genes for ear size, ability to roll the tongue and eye colour are all located on the same chromosome – that is they are linked and are usually inherited together. These are passed on as a group and are called **linked alleles**.

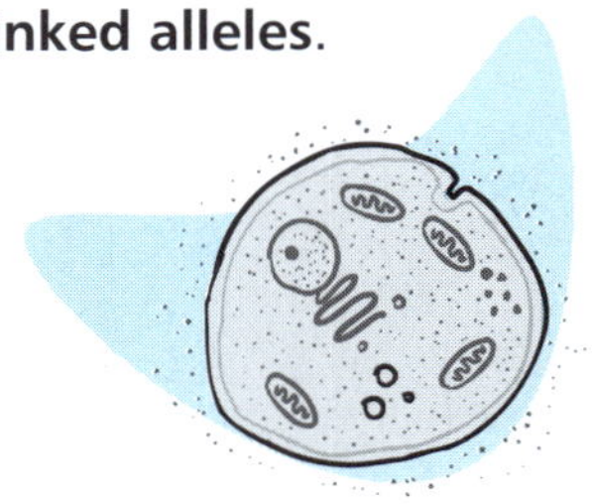

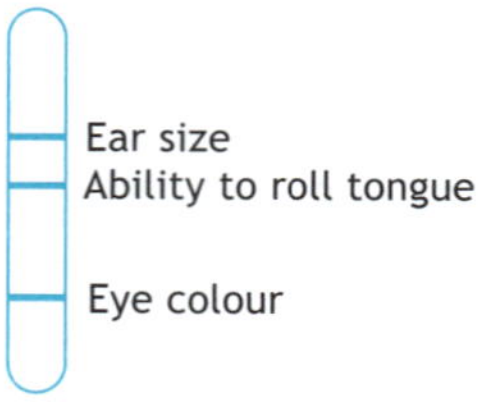

Segment of a chromosome showing linked alleles

- Sex chromosomes contain many alleles, including the ones that determine gender. The Y chromosome is very short, so in males all the genes on the X chromosome, whether they are recessive or dominant, are expressed. These alleles which are carried by the sex chromosomes are called **sex-linked alleles**.

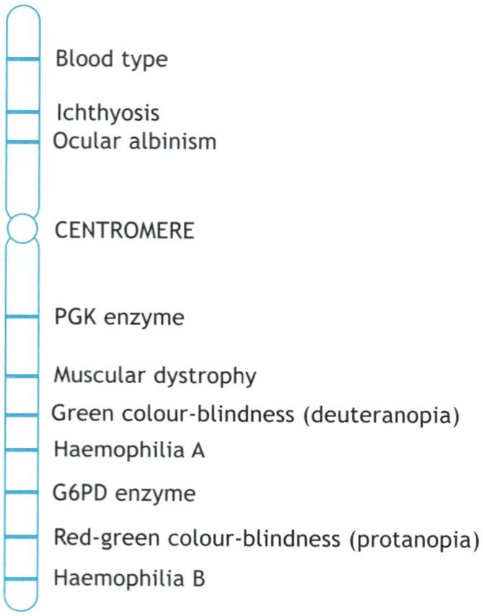

Linkage on the human X chromosome

Mutation

Genes or alleles are reshuffled by meiosis, but accidental changes can occur. This happens through a rearrangement in the DNA molecule or a change in the shape or number of chromosomes.

If mutations occur in the sex cells then the change is passed on. Mutations are a source of variation and allow evolution to proceed.

Genetic engineering

In genetic engineering, scientists deliberately change genotypes by moving genes from one organism to another. Genetic engineers can therefore give an organism characteristics that it does not normally have.

Genetic engineering procedures can be used in many ways, including the identification of genes; the production of proteins for medical use; the inserting of functional genes to correct disorders; the insertion of genes into livestock and plants to improve quality.

The Human Genome Project, which listed all 3.1 billion bases in the human DNA, was completed in 2000. These procedures are raising serious ethical issues; however, providing it is used responsibly, the tool of gene technology is promising a very interesting future.

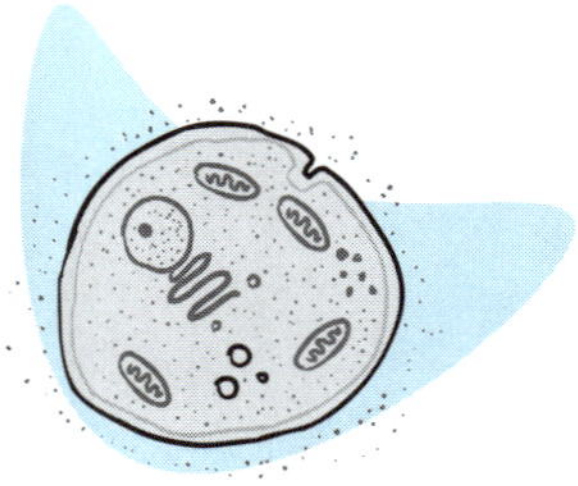

Glossary

Allele – One of two or more forms of the same gene.

Amino acid – The building blocks of proteins that can be arranged in a vast number of sequences.

Base – A substance that forms a store of information in a nucleic acid.

Base pairs – Two bases linked chemically.

Chromosome – A thread-like structure that contains all or part of a cell's DNA.

Codon – A three-base group used in the genetic code.

DNA – A substance that stores information in all living things.

Dominant allele – An allele that usually shows itself in a living thing's phenotype.

Double helix – The DNA molecule has a spiral shape and two strands linked by huge numbers of base pairs.

Evolution – A gradual genetic change in a group of living things.

Gene – A basic unit of heredity that carries the instructions needed to make a particular protein.

Genetic code – The chemical code used by DNA and RNA.

Genome – All the information encoded in our genes.

Genotype – The genetic makeup of a cell or living thing.

Heterozygous – A cell with pairs of different alleles.

Homozygous – A cell with pairs of the same alleles.

Linked alleles – Alleles that tend to be inherited together.

Locus – A position on a chromosome.

Mutation – A change in a cell's genetic material.

Nucleic acid – Complex organic compounds called RNA and DNA that are found in all living things and carry information.

Nucleotide – One of the molecules that add together to make a nucleic acid.

Phenotype – The physical characteristics of an organism.

Protein – Large molecules made up of hundreds of units called amino acids.

Replication – The process by which a nucleic acid molecule copies itself.

Ribosome – A small body made of RNA and protein. It is the site of protein synthesis and the process of translation. Cells contain many thousands of ribosomes, which are found in the endoplasmic reticulum of the cytoplasm.

RNA – A group of nucleic acids that handle the information used to make proteins.

Recessive allele – An allele that is masked by a dominant allele.

Sex-linked alleles – An allele that is carried by a sex chromosome.

Variation – The natural difference between living things.

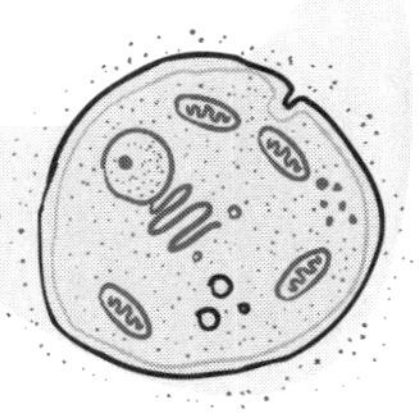

Science skills

Question 1

Study the diagram of a double helix below.

a Label the three main parts.

b Write the names that the letters C, G, A, T stand for.

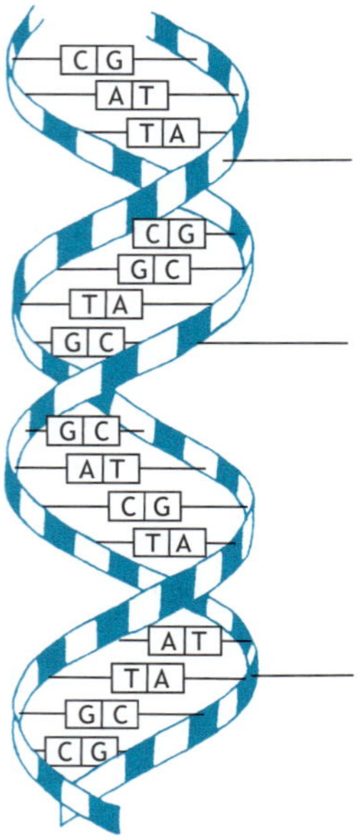

Question 2

When a cell divides, its DNA molecules copy themselves in order to pass on a set of instructions to each new cell.

a What is the name of this process?

b Arrange the following sentences in the correct order to describe this process.

- Two new DNA molecules are produced.
- The two strands form a complementary copy of themselves using free bases in the cell.
- DNA unzips itself to form two separate strands.

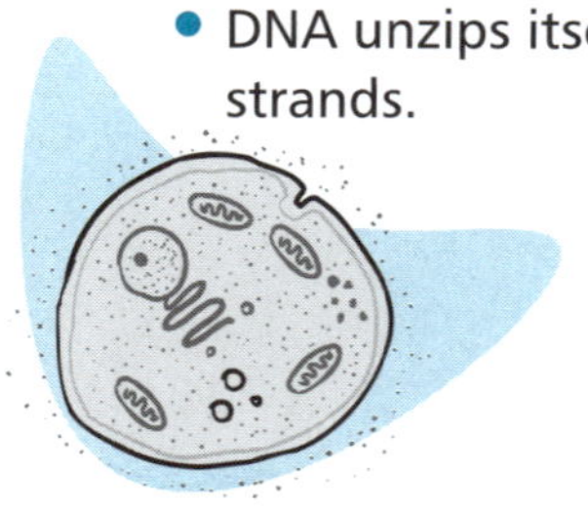

c The sequence of bases on one strand of a DNA molecule is C-G-G-A-T-T. What is the sequence on the complementary strand of DNA?

Question 3

Complete the diagram of a section of DNA by adding the names of the missing bases.

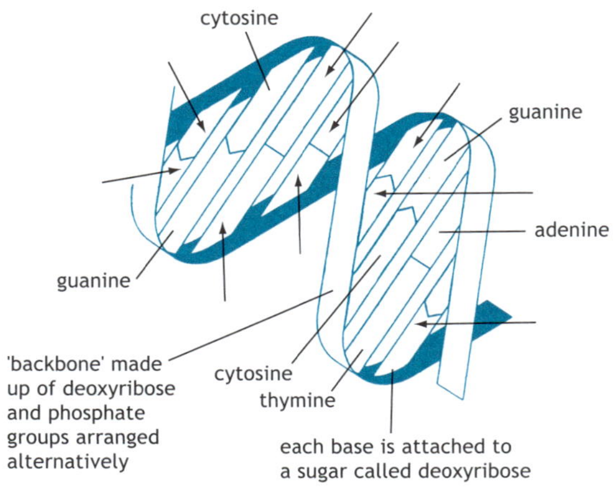

Question 4

a In a cell, where exactly are the instructions located for making proteins?

b The following diagram represents the first stage of protein formation. What is it called?

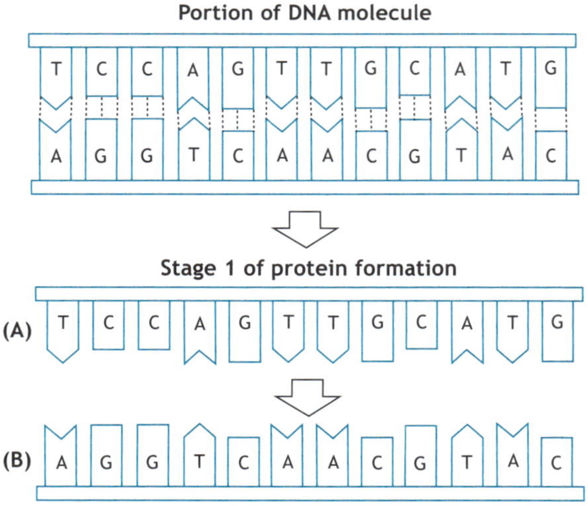

c Is strand 'A' a strand of DNA or RNA? ________

d Is strand 'B' a strand of DNA or RNA? ________

e What is the second stage of protein synthesis called?

f Complete the following description of the second stage of protein synthesis.

Chemical clusters called __________ move along the new strand of __________ RNA, reading the strand's instructions one codon at a time. With the help of __________ RNA, the ribosomes convert the sequence of __________ into the required sequence of __________ __________, thus forming a new __________.

Question 5

Mendel experimented with flower colour by crossing red flowers and white flowers.

a What is the phenotype of each of the flowers?

____

b What is the genotype of the red flowers in the first generation?

____

c What are the alleles of these red flowers?

____

d What is the ratio of red flowers to white flowers in the second generation?

____

Question 6

Another of Mendel's experiments involved peas with smooth skins and some with wrinkled skins. The gene for smooth skins is dominant (S) and both types of peas were homologous.

a The smooth skinned peas were crossed with the wrinkled skinned peas. What was the genotype each parent?

____

b What was the genotype of the first generation?

____

c What was the phenotype of the first generation?

____

d What was the ratio of smooth skins to wrinkled skins in the second generation?

____

e What were the genotypes of the 3 smooth skinned peas in the second generation?

____

Question 7

Geneticists have discovered that the allele which produces long floppy ears in rabbits is dominant over the allele for short ears. A homozygous long-eared male rabbit is mated with a female rabbit with short ears.

a What is the genotype of the male and the female?

____

b What will be the genotype and phenotype of the first generation?

____

____

Question 8

A male and a female rabbit from the first generation above were mated.

a What are the possible genotypes in the second generation?

____

b What are the possible phenotypes in the second generation?

____

c In what ratios are the genotypes found?

____

Question 9

A small proportion of a cat population has hair that stands up along the backbone. This is caused by a recessive gene. Individuals with this characteristic are referred to as ridgebacks. To increase the frequency of this trait a breeding program was set up.

Three different matings were set up:

i homologous normal fur male x ridgeback female

ii heterozygous normal fur male x ridgeback female

iii heterozygous normal fur male x heterozygous normal fur female

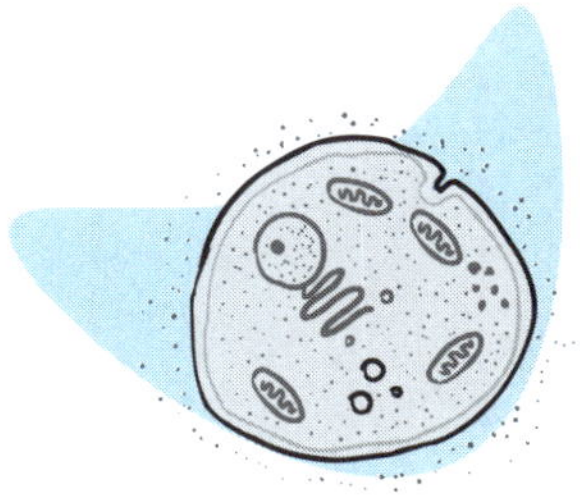

a What are the genotypes of each of the parents in each of the matings?

b What are the possible phenotypes, genotypes and their ratios in each of the matings?

c Which of the three matings is likely to increase the number of ridgebacks in the population quickly? Explain your answer.

Question 10

A homogeneous bull, dominant for red colour, is mated to a homogeneous white cow.

Draw a plan of the breeding program. Include the two parents, the first generation and the second generation.

Question 11

Which of the following matches is incorrect?

a Heterozygous – A cell with pairs of different alleles.

b Recessive gene – The gene which is known to be present in a studied individual, though its effect is suppressed.

c Phenotype – The composition of the genetic material in a species.

d Dominant gene – A gene which produces its effect even though another allele is present.

Question 12

Study the plan of eye colour inheritance in the diagram below.

a the offspring have blue eyes or brown eyes?

b Is it possible for this offspring to have a blue-eyed child?

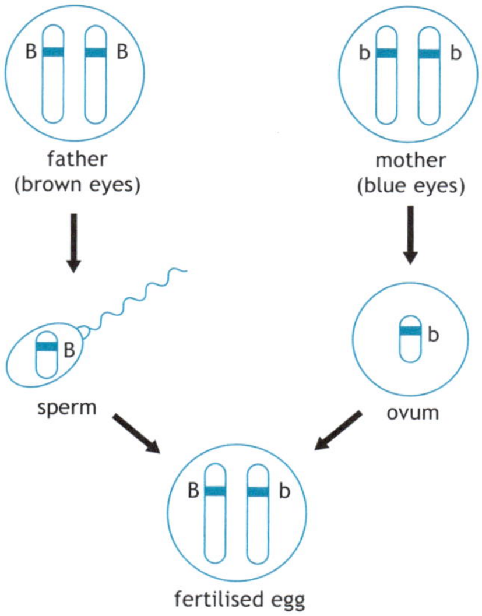

Question 13

Alleles which are carried by the sex chromosomes are called **sex-linked alleles**. One characteristic that is carried by a gene on the 'X' sex chromosome is called haemophilia. If a male has the recessive gene on his X chromosome, he is a haemophiliac, but a female must carry a recessive gene on both X chromosomes to actually have the disease.

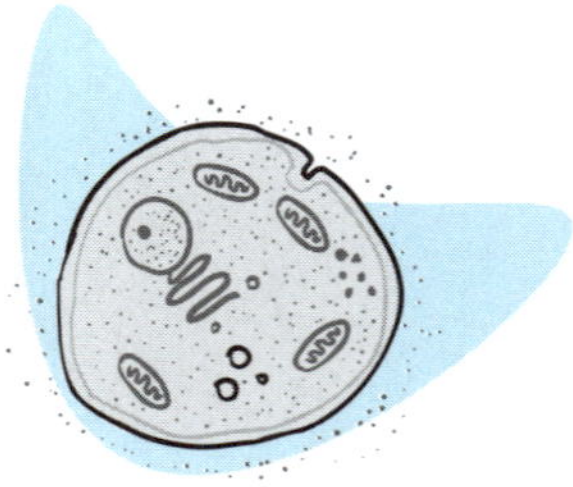

The figure below shows the family tree of Queen Victoria. Which females would have to have at least one recessive gene to explain this result?

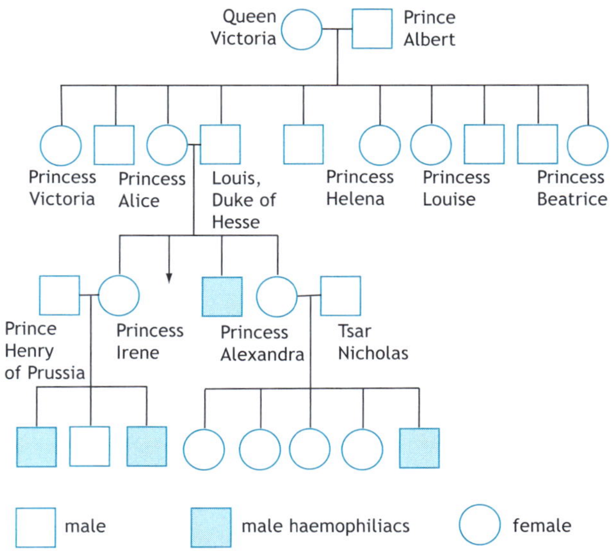

Question 14

A team of biotechnologists have been working on mapping the genetic code of wolves in the French Alps. They would like to know where these wolves originally came from as well as keeping track of each wolf and understanding more about their way of life. During their research they have worked out the bases of part of the DNA sequence of a wolf, a wild dog and a domestic dog.

	Wolf	Wild dog	Dog
120	ACGTCGTGCA	ACGTCGTGCA	ACGTCGTGCA
130	TTAATGGTTT	TTAATGGTTT	TTAATGGTTT
140	GCCCCATGCA	GCCCCATGCA	GCCCCATGCA
150	TATAAGCATG	TATAAGCATG	TATAAGCATG
160	TACATAATAT	TACATAATAT	TACATAATAT
170	TATATCCTTA	TATATCCTTA	TATATCCTTA
180	CATAGGACAT	CATAGGACAT	CATAGGACAT
190	ACTAACTCAA	ACTAACTCAA	A TAACTCAA
200	TCTCACAATT	TCTCACAATT	TCTCA AATT
210	CACTGACCTA	CACTGACCTA	CACTGA CT
220	TCAACAGTAA	TCAACAGTAA	TCAACAGTAA

Based on this information, is the wolf closest to the wild dog or the domestic dog?

Question 15

Which of the following applications of gene technology are likely in the future?

a The identification and mapping of genes of many species.

b The production of proteins for medical use.

c The inserting of functional genes to correct diseases and disorders.

d The insertion of genes into livestock and plants to improve quality.

e The creation of a real Jurassic Park.

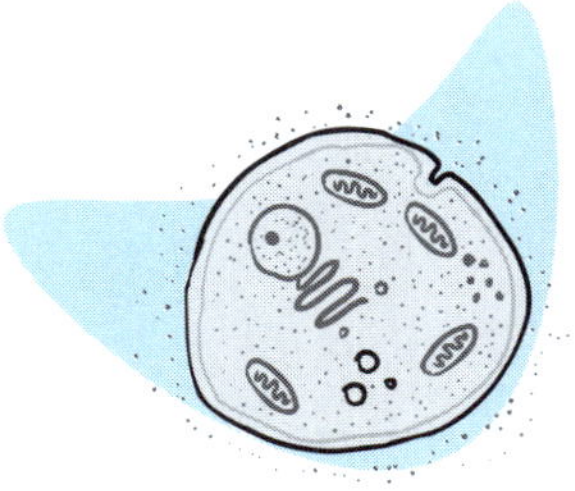

Revision questions

Question 1

Match the following terms with the correct explanation.

Codon	A thread-like structure that contains all or part of a cell's DNA.
DNA	A three-base group used in the genetic code.
Chromosome	A basic unit of heredity that carries the instructions needed to make a particular protein.
Gene	Deoxyribonucleic acid.

Question 2

What name is given to the following process?

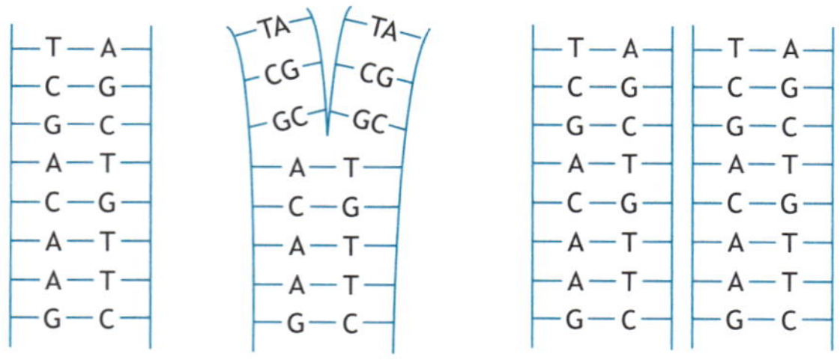

Chromosomes

Question 3

In a DNA molecule two threads are wound around each other as shown in the diagram below.

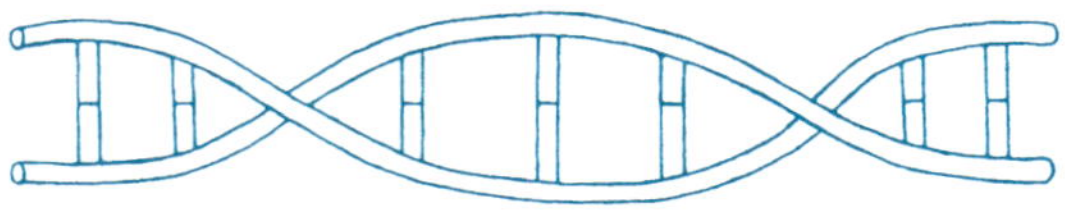

a What is this structure called?

____

b What holds the strands together?

____

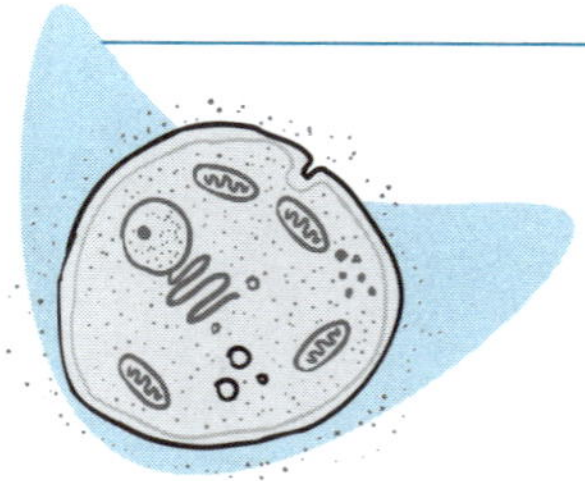

c What process is RNA involved in?

____

d Name the two forms of RNA.

____

____

Question 4

a Which famous scientists worked out the structure of DNA?

____

b Which scientist worked out the mathematical rules for inheritance?

____

Question 5

Add 'RNA' or 'DNA' to complete the sentences below:

_______ is mostly found in the cell nucleus, but _______ is found in the cell nucleus and the cell cytoplasm. _______ contains the information for making proteins. _______ carries this information to the ribosomes in the cytoplasm. _______ is the substance which determines what the organism will look like.

Question 6

a What are the three basic chemical components that fit together to form a DNA molecule?

____

____

____

b What is the relationship between a gene and a chromosome?

____

____

____

c What is the basic chemical structure of a gene?

____

____

____

____

Question 7

a What is the genetic code?

____

____

b What are the four nitrogen-containing bases that make up DNA?

____

____

c What does the genetic code tell the cell?

____

____

d A sequence of three bases or a 'triplet' forms an __________ __________ .

Question 8

Write the correct term to fit the descriptions below.

a The visible characteristics of an organism.

____

b The link between successive generations of living things.

____

c One of two or more forms of the same gene.

____

d The genetic make up of a cell or living thing.

____

Question 9

a When genes appear in pairs on a chromosome, as in the diagram below, what are they called?

____

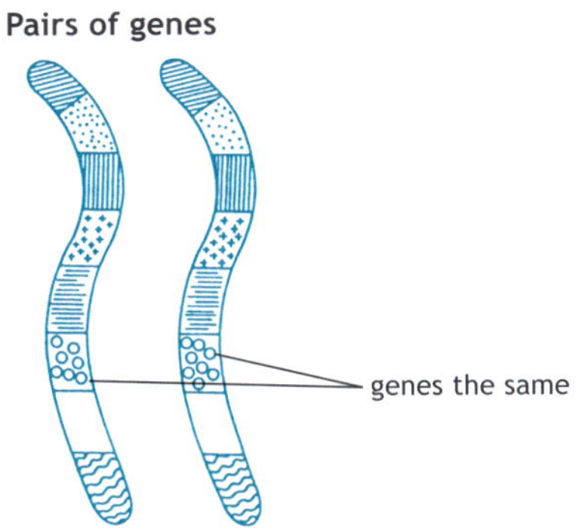

Question 10

Two parents of the same species produce offspring that are like themselves but not exactly the same. This difference is described as 'variation'.

a What process depends on variation within a species?

____

b What is another way that variation within a species occurs?

____

Question 11

The following mice are the offspring of a black father and a white mother.

a What was the genotype of the father?

____

b What was the genotype of the mother?

____

c What is the phenotype of the offspring?

____

d Were the parents homozygous or heterozygous?

____

e Are the offspring homozygous or heterozygous?

____

Question 12

Mendel crossed plants of differing heights. The parents were heterozygous for the height gene. Complete the expected results below.

a Parent genotypes:

____

b Possible alleles in each parent:

____

c Possible offspring genotypes:

____

d Possible offspring phenotypes:

____

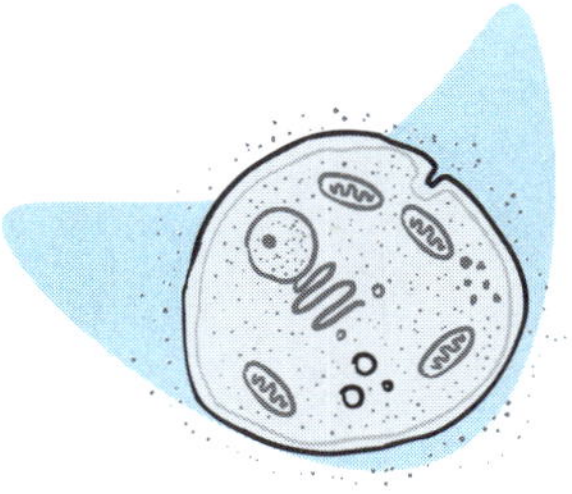

Question 13

Match the following terms with the correct descriptions:

Homozygous cell	Alleles carried by the sex chromosomes.
Heterozygous cell	A change in a cells genetic material.
Linked alleles	A cell with identical alleles.
Sex-linked alleles	Alleles that are close to each other on the chromosome and are inherited together.
Mutation	A cell with one dominant and one recessive allele.

Question 14

Study the diagram below of two homologous chromosomes.

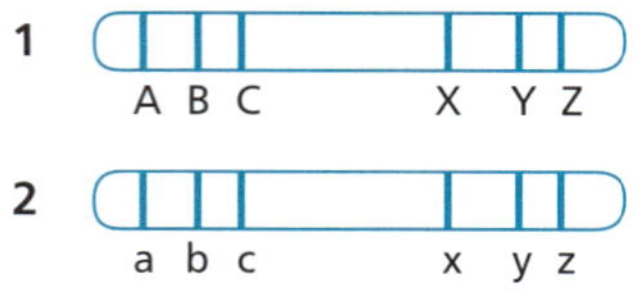

Write the missing word in the following sentence:

Alleles A, B, C, X, Y, Z occupy the same __________ on chromosome 1 as alleles a, b, c, x, y, z on chromosome 2.

Question 15

Complete the information below by adding the correct words.

The information encoded in human genes is called the H __________ G __________. This information is written in a code involving four bases: __________, __________, __________ and __________. These bases are strung together by material made of and . This information spells out how to make life.

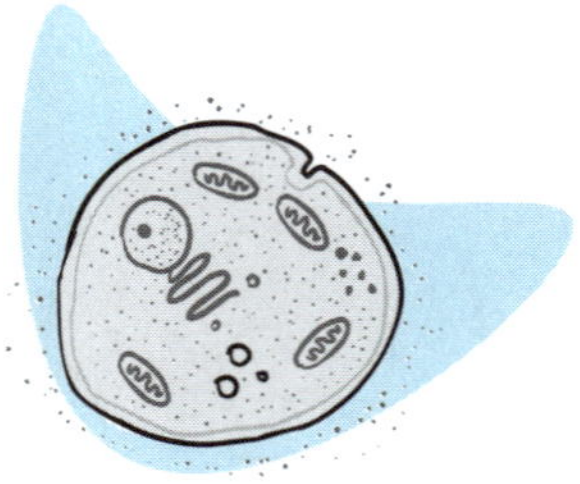

CHAPTER 3

ATOMIC STRUCTURE AND THE PERIODIC TABLE

What is the language of chemistry?

Everything in the world can be organised into two concepts: matter and energy. Matter is anything that takes up space and has **mass**. Matter is composed of **atoms**. Over 100 different types of atoms have been discovered. These chemicals are called elements, each with its own kind of atom and unique character. These elements combine in different ways to make millions of substances.

The structure of an atom

The atoms of all matter are electrical and are made up of three small subatomic particles:

- **Electrons**, which carry negative charges.
- **Protons**, which carry positive charges.
- **Neutrons**, which carry no charge.

Protons and neutrons make up the very dense nucleus of the atom and the tiny electrons are arranged in shells of different energy levels and orbit around the nucleus. The maximum number of electrons in the first shell is always two, the second is eight and the third is eight. Electrons fill the inside shells first.

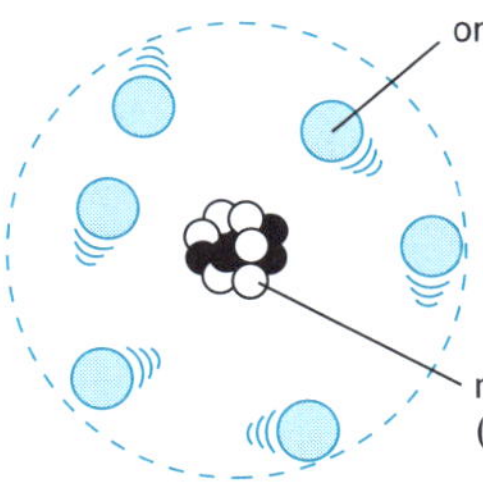

Basic structure of an atom (solar system model)

The Solar System Model of the atom was proposed by Niels Bohr. This model describes an atom as mostly empty space, but containing very small particles called electrons, which seem to orbit the dense and positively charged nucleus like the planets orbiting the Sun.

The electric charge on an electron is exactly equal and opposite to that on a proton. The electric charge of a proton is exactly equal and opposite to that of an electron. An atom therefore has no overall charge.

Atoms are held together by a strong attraction between the protons and electrons. Particles with opposite charges attract one another, like opposite poles of a magnet. For example:

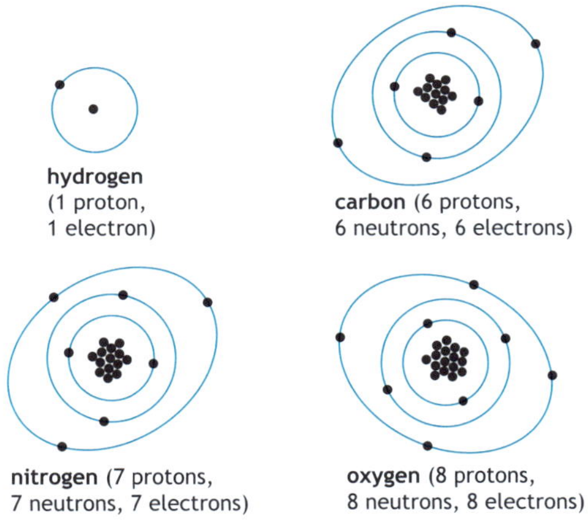

The four most common atoms in biological compounds

Molecules

Some atoms exist alone but most join others to form molecules. **Compounds** are made of identical molecules, each with the same combination of atoms. For example a glucose molecule has 6 carbon atoms, 12 hydrogen atoms and 6 oxygen atoms, and all water molecules consist of 2 hydrogen atoms and 1 oxygen atom.

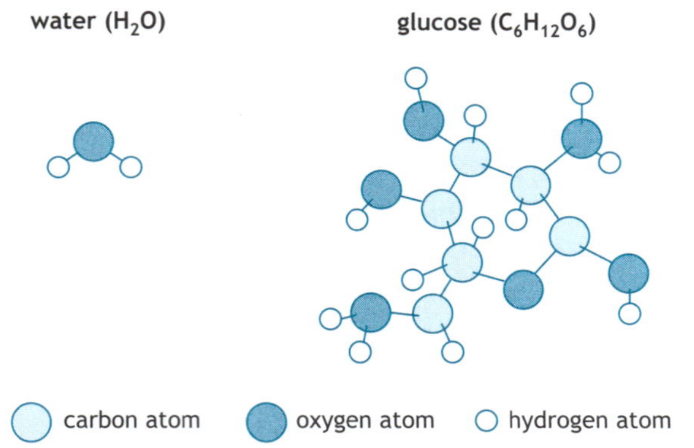

Water molecule **Glucose molecule**

Ions

The addition or removal of electrons from atoms or groups of atoms creates ions. An atom that has lost one or more of its electrons is positively charged and is called a **cation**. An atom that has gained one or more electrons is called an **anion** and is negatively charged. Interestingly, a hydrogen ion is a proton, as hydrogen atoms consist of only one proton.

An atom with only 1 or 2 electrons in its outer shell tends to lose them, but an atom with 6 or 7 electrons tends to gain enough electrons to fill the outer shell.

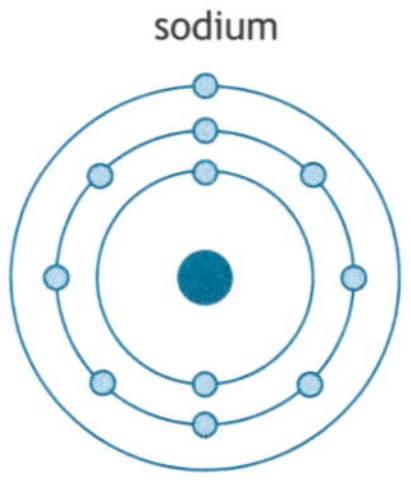

Sodium prefers to lose an electron – it becomes positively charged (Na+)_

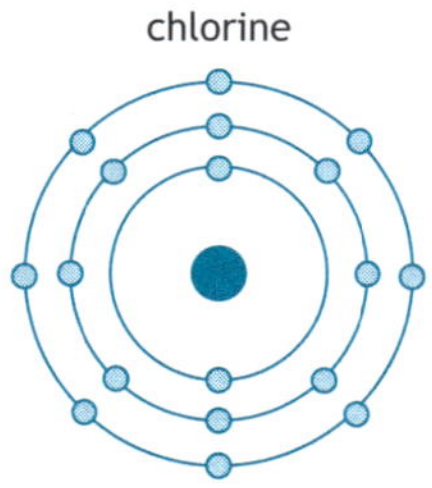

Chlorine prefers to gain an electron – it becomes negatively charged (Cl–)

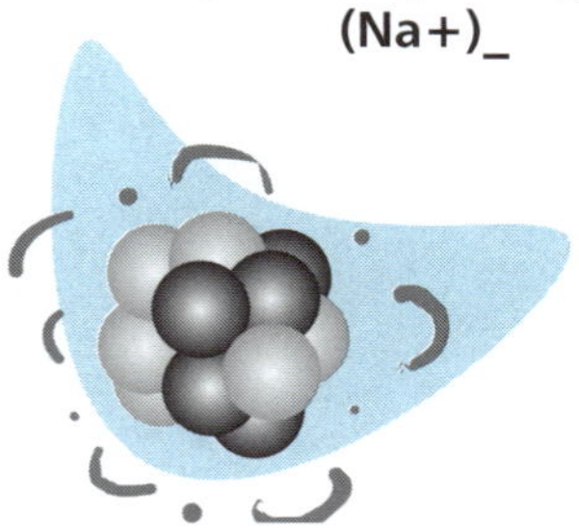

The charge on the electron is equal to the number of electrons gained or lost.

For example, each calcium (Ca++) has lost two electrons and each iodide (I–) has gained one electron.

How do atoms and substances differ?

Why is gold different from carbon? Elements can be incredibly different from one another. Each element is composed of different types of atoms and each atom is composed of different numbers of subatomic particles – protons, neutrons and electrons.

For example, the table below shows the four simplest elements:

Element	Protons	Electrons	Neutrons
hydrogen	1	1	0
helium	2	2	2
lithium	3	3	4
beryllium	4	4	5

Hydrogen contains only one proton and one electron and no other element is the same as this. No two elements have atoms with the same number of protons.

Atoms are neutral and the size of the charge on protons and electrons is the same, but opposite in sign. To have a neutral atom the number of electrons in an atom must be the same as the number of protons. The number of protons is important in defining variation in atoms.

- The **atomic number** of an element is the number of protons in the nucleus of an atom of that element. Hydrogen has one proton so it has an atomic number of one.
- The **mass number** is the total number of protons and neutrons in the nucleus of an atom. For example sodium has an atomic number of 11 and a mass number of 23. Sodium therefore has 11 protons, 11 electrons and 12 neutrons. Mass numbers are not unique to one particular atom, but atomic numbers are.

- The **atomic weight** of an element is the average mass of naturally occurring atoms of an element.

So atoms are different because they have different numbers of protons, neutrons and electrons. The numbers of neutrons and protons cannot change in a chemical reaction, however electrons can be added to or subtracted from an atom.

To summarise:

Atoms differ depending on the type and number of subatomic particles they contain as well as the patterns in which the particles are combined. Substances differ depending on which elements are present, how many atoms are present and the patterns in which the atoms of the elements are combined.

Naming elements

Each element or type of atom has a name as well as a unique chemical symbol of one or two letters. This simplifies things when referring to the atoms in **molecules** and **compounds**. The letters are either:

- the first letter of the element's name – for example H for hydrogen and O for oxygen
- two letters from the element's name – for example Ca for calcium, Cr for chromium, or
- from the Latin name – for example Au for gold, Fe for iron.

Examples of compound formula:

Element	Symbol
Hydrogen	H
Carbon	C
Nitrogen	N
Sodium	Na
Magnesium	Mg
Potassium	K
Zinc	Zn

Organising the elements

As more and more elements were discovered, chemists looked for trends in the properties of elements. The elements were arranged in order of increasing atomic number across a grid. This grid was the most significant development in chemistry of the nineteenth century and is called the periodic table (see next page).

The periodic table now contains over 100 elements. These are arranged in:

- rows, or **periods**, numbered 1 to 7 down the table, of increasing atomic number, and
- columns, or **groups**, labelled 1 to 18, which contain up to seven elements, all with similar properties. The element with the smallest atomic number is at the top of the column.

What is the periodic table used for?

Currently, the periodic table is used for two main purposes:

- as a summary of atomic weights and numbers, symbols and names for the elements
- to help in knowing the properties of elements.

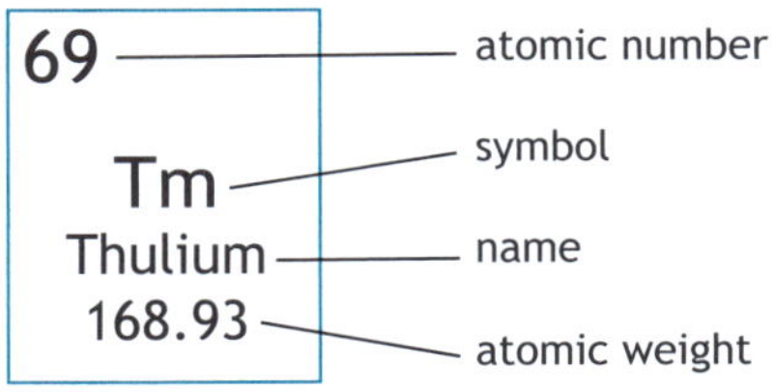

Element Information Contained in the Periodic Table

PERIODIC TABLE OF THE ELEMENTS

KEY

Atomic Number	79	Symbol of element
	Au	
Atomic Weight	197.0	Name of element
	Gold	

1	2	3	4	5	6	7	8	9	10	11	12	13	14	15	16	17	18
1 H 1.008 Hydrogen																	2 He 4.003 Helium
3 Li 6.941 Lithium	4 Be 9.012 Beryllium											5 B 10.81 Boron	6 C 12.01 Carbon	7 N 14.01 Nitrogen	8 O 16.00 Oxygen	9 F 19.00 Fluorine	10 Ne 20.18 Neon
11 Na 22.99 Sodium	12 Mg 24.31 Magnesium											13 Al 26.98 Aluminium	14 Si 28.09 Silicon	15 P 30.97 Phosphorus	16 S 32.07 Sulfur	17 Cl 35.45 Chlorine	18 Ar 39.95 Argon
19 K 39.10 Potassium	20 Ca 40.08 Calcium	21 Sc 44.96 Scandium	22 Ti 47.87 Titanium	23 V 50.94 Vanadium	24 Cr 52.00 Chromium	25 Mn 54.94 Manganese	26 Fe 55.85 Iron	27 Co 58.93 Cobalt	28 Ni 58.69 Nickel	29 Cu 63.55 Copper	30 Zn 65.38 Zinc	31 Ga 69.72 Gallium	32 Ge 72.64 Germanium	33 As 74.92 Arsenic	34 Se 78.96 Selenium	35 Br 79.90 Bromine	36 Kr 83.80 Krypton
37 Rb 85.47 Rubidium	38 Sr 87.62 Strontium	39 Y 88.91 Yttrium	40 Zr 91.22 Zirconium	41 Nb 92.91 Niobium	42 Mo 95.96 Molybdenum	43 Tc Technetium	44 Ru 101.1 Ruthenium	45 Rh 102.9 Rhodium	46 Pd 106.4 Palladium	47 Ag 107.9 Silver	48 Cd 112.4 Cadmium	49 In 114.8 Indium	50 Sn 118.7 Tin	51 Sb 121.8 Antimony	52 Te 127.6 Tellurium	53 I 126.9 Iodine	54 Xe 131.3 Xenon
55 Cs 132.9 Caesium	56 Ba 137.3 Barium	57–71 Lanthanoids	72 Hf 178.5 Hafnium	73 Ta 180.9 Tantalum	74 W 183.84 Tungsten	75 Re 186.2 Rhenium	76 Os 190.2 Osmium	77 Ir 192.2 Iridium	78 Pt 195.1 Platinum	79 Au 197.0 Gold	80 Hg 200.6 Mercury	81 Tl 204.4 Thallium	82 Pb 207.2 Lead	83 Bi 209.0 Bismuth	84 Po Polonium	85 At Astatine	86 Rn Radon
87 Fr Francium	88 Ra Radium	89–103 Actinoids	104 Rf Rutherfordium	105 Db Dubnium	106 Sg Seaborgium	107 Bh Bohrium	108 Hs Hassium	109 Mt Meitnerium	110 Ds Darmstadtium	111 Rg Roentgenium	112 Cn Copernicium		114 Fl Flerovium		116 Lv Livermorium		

Lanthanoids

57 La 138.9 Lanthanum	58 Ce 140.1 Cerium	59 Pr 140.9 Praseodymium	60 Nd 144.2 Neodymium	61 Pm Promethium	62 Sm 150.4 Samarium	63 Eu 152.0 Europium	64 Gd 157.3 Gadolinium	65 Tb 158.9 Terbium	66 Dy 162.5 Dysprosium	67 Ho 164.9 Holmium	68 Er 167.3 Erbium	69 Tm 168.9 Thulium	70 Yb 173.05 Ytterbium	71 Lu 175.0 Lutetium

Actinoids

89 Ac Actinium	90 Th 232.0 Thorium	91 Pa 231.0 Protactinium	92 U 238.0 Uranium	93 Np Neptunium	94 Pu Plutonium	95 Am Americium	96 Cm Curium	97 Bk Berkelium	98 Cf Californium	99 Es Einsteinium	100 Fm Fermium	101 Md Mendelevium	102 No Nobelium	103 Lr Lawrencium

Finding your way around the periodic table

- Those elements with only one electron in their outer shell, such as lithium, sodium and potassium, are called the alkali metals (group 1) and combine easily with other elements. They are the most reactive elements.
- Similarly those elements which need just one more electron to complete their outer shells, such as fluorine, chlorine and bromine, are alike in character and are called halogens (group 17). These are also very reactive elements.
- Some elements have six electrons in the outer shell (group 16) and so need two more electrons to fill it. These are, therefore, less reactive elements.
- The zigzag line separates the metals from the non-metals. All the elements on the left-hand side of the line are metals; all those on the right-hand side are non-metals. Metals have high lustre, conduct electricity and are malleable. Non-metals are non-lustrous and poor conductors.
- Metals in groups 1, 2 and 3 lose electrons when they form ions. The non-metals in groups 15, 16 and 17 all gain electrons when they form ions.
- The transition elements (columns 3–12) are found in the middle and in the separate grid. These elements do not fit into one of the main eight groups, as they have more complex electron arrangements.
- Some elements, such as helium, neon, argon and krypton, have their outer shells completely full. This makes them very stable and slow to react with other elements (column 18). These are inert and are called the noble gases. They glow brightly when an electric current is passed through them.

By looking at an element's position in the periodic table, scientists can learn a great deal about how it will behave and they can also predict how elements yet to be discovered will behave.

Glossary

Anion – A negatively charged ion.

Atom – The smallest unit of a chemical element. Atoms contain electrons, protons and neutrons.

Atomic number – The number of protons in an atom. Atoms are electrically neutral so the number of protons in the nucleus of an atom must equal the number of electrons its nucleus.

Atomic weight – The average mass of naturally occurring atoms of an element.

Cation – A positively charged ion.

Compound – A substance that consists of more than one kind of atom.

Electron – A particle with a negative electric charge, found in atoms. The electric charge on an electron is exactly equal and opposite to that on a proton. The addition or removal of electrons from atoms creates ions.

Element – A substance consisting of atoms all of the same kind. One element cannot be changed into another element except by splitting atoms.

Group – A column of elements in the periodic table.

Ion – An atom or molecule with an electric charge, due to having an unequal number of protons and electrons. An ion with a positive charge has more protons than electrons, and an ion with a negative charge has fewer protons than electrons.

Mass – The amount of matter that an object contains. The standard unit of mass is the kilogram.

Mass number – The combined total of protons and neutrons in a nucleus.

Molecule – The smallest unit of an element or compound. Molecules mostly consist of more than one atom.

Neutron – A particle with no electric charge, found in all atoms. Neutrons have the same weight as protons.

Period – A row of elements in the periodic table.

Periodic table – An arrangement of the known elements by ascending atomic number and by similar properties.

Proton – A particle with a positive electric charge found in all atoms. The electric charge of a proton is exactly equal and opposite to that of an electron. An atom therefore has no charge.

Science skills

Question 1

The diagram below is of an atom.

a Label the diagram with the terms electron, proton, neutron and nucleus.

b What is the name of the model used to draw this atom?

____

c Who proposed the model?

____

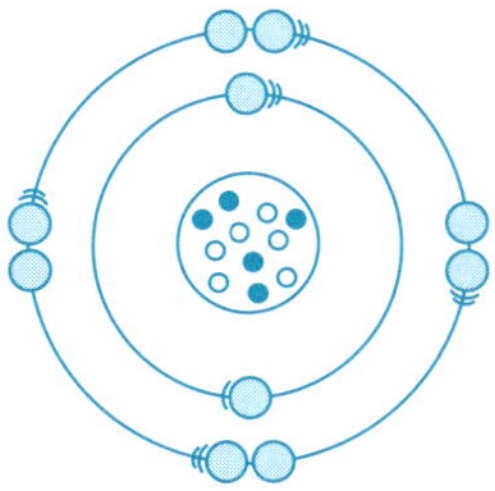

Question 2

Complete the following three equations.

a atomic number = number of ________

= number of ________

b _____ _______ = number of _________ + number of neutrons

c number of neutrons = _____ number – ________ number

Question 3

Write the names of the elements beside the symbols shown over page.

Element	Symbol
	H
	C
	Na
	Zn
	Cl
	O

Question 4

What happens to an atom of hydrogen if it:

a gains an electron?

b loses an electron?

c Complete the table below.

Case	Protons	Electrons	Overall charge
neutral atom	1		
gain an electron	1		
lose an electron	1		

Question 5

a Draw the electron orbit arrangement for hydrogen, helium, lithium and sodium.

b Write the chemical symbols for each.

c Which ones are the most reactive?

d Why are they the most reactive?

Question 6

a Complete the table below.

Substance	How many types of atoms	Element or molecular compound
oxygen		
water		
glucose		
hydrogen		

b Draw a diagram of the atomic structure of oxygen and water molecules.

Question 7

Study the following key and then answer the questions below.

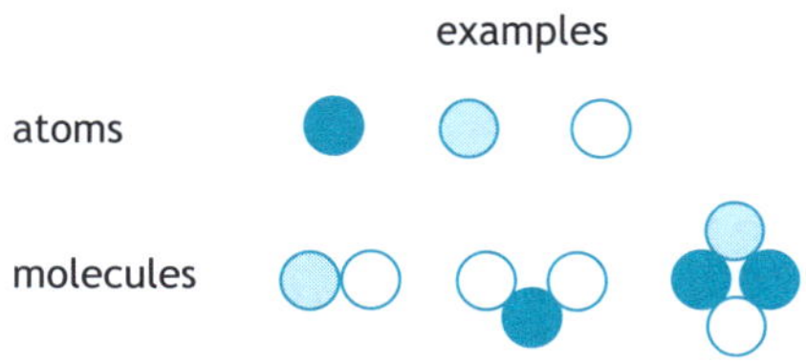

a Which of the following substances are elements?

b Which of the following substances are the molecules of compounds? Why?

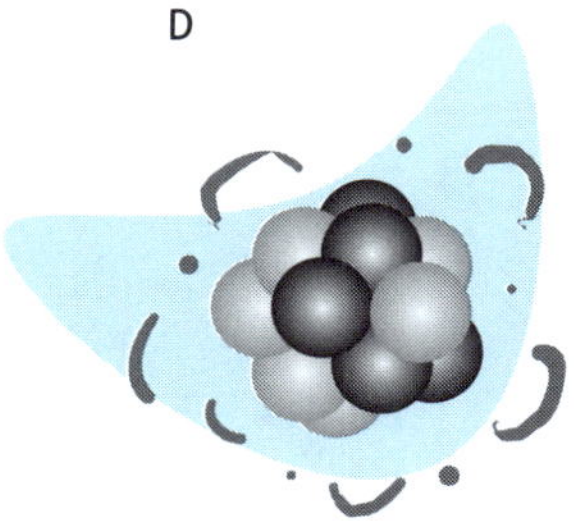

Question 8

Complete the table by adding the full name of the element.

Symbol	Element
H	
He	
Li	
C	
N	
Na	
Zn	
Ar	
Fe	

Question 9

The following diagram is part of the periodic table.

1	2	3	4	5	6	7	8	9	10	11	12	13	14	15	16	17	18
1 H 1																	2 He 4
3 Li 7	4 Be 9											5 B 11	6 C 12	7 N 14	8 16	9 F 19	10 Ne 20
11 Na 23	12 Mg 24											13 Al 27	14 Si 28	15 31	16 S 32	17 Cl 35.5	18 40
19 K 39	20 Ca 40	21 Sc	22 Ti	23 V	24 Cr	25 Mn	26 Fe 56	27 Co	28 Ni	29 Cu 63.5	30 Zn 65	31 Ga	32 Ge	33 As	34 Se	35 Br 80	36 Kr

a Write in the missing elements.

b In which column are the noble gases found?

c In which column would you find the elements with only one electron in the outer shell?

d In which column are the elements that need one more electron to complete their outer shell?

e Does column 6A or column 7A have the most reactive substances? Why?

f Which elements are non-metals?

g What are the symbols of the transition elements?

h In which column are the halogens found?

Question 10

a The following diagram is just one of the boxes from the periodic table. Label the four pieces of information.

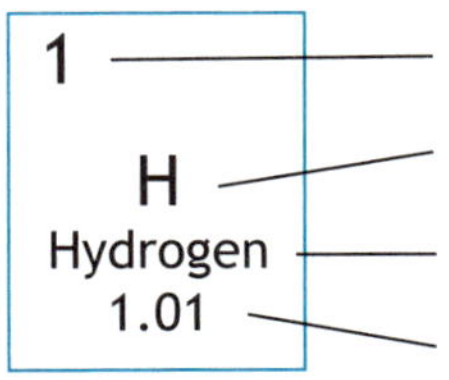

b Find the element Calcium (Ca) on the periodic table. What is its atomic number and atomic weight?

Question 11

Use the periodic table to complete the following information.

Element	Likely electron movement	Does it become a cation or anion?
sulphur		
magnesium		
calcium		
argon		
bromine		

Question 12

Part of the periodic table is provided below. It shows the symbol and the atomic number of the elements.

a Study the key and label with the correct sections of the periodic table.

b Write the atomic number for sodium (Na), oxygen (O) and copper (Cu).

____

c Which elements have the atomic numbers of 17, 6 and 11?

____

d How many electrons are in the outer orbit of magnesium (Mg), neon (Ne) and nitrogen (N)?

____

Question 13

Scientists have been studying the atom for a long time. One experiment that has been carried out was to find the amount of energy required to remove one electron from an atom of an element. The graph of this data is presented below.

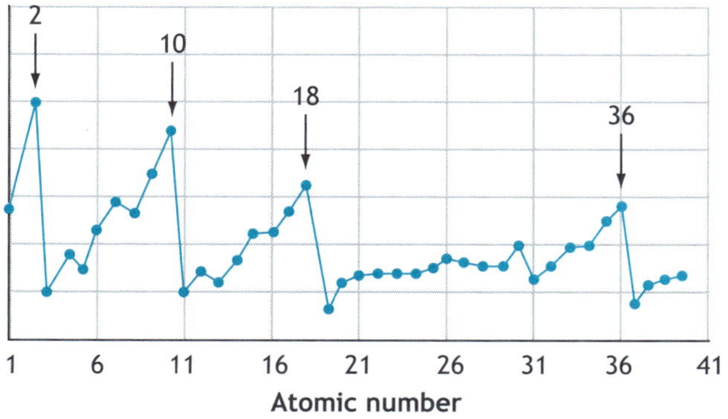

a Which elements need the most energy to remove an electron?

____

b Which group of elements do they belong to?

____

c Why do they take so much energy to remove an electron?

____

____

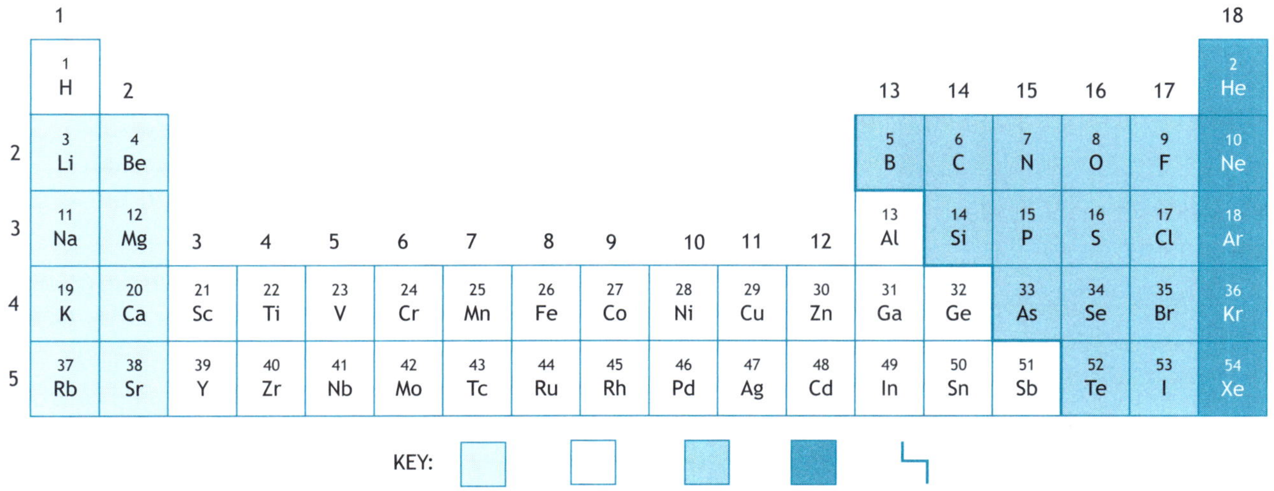

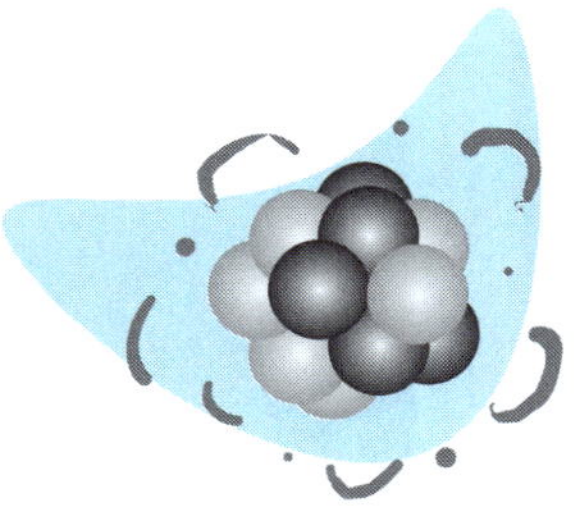

Question 14

There are several ways to classify elements. Examples below are based on physical state, metallic character and origin.

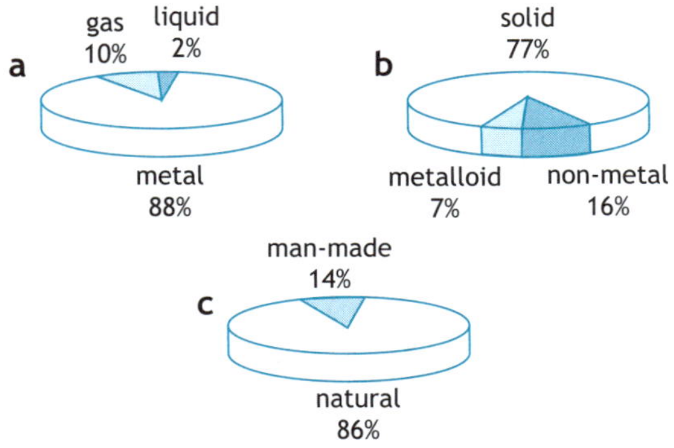

Which classification system was used to construct the periodic table?

Question 15

Study the following Solar System Model of an atom.

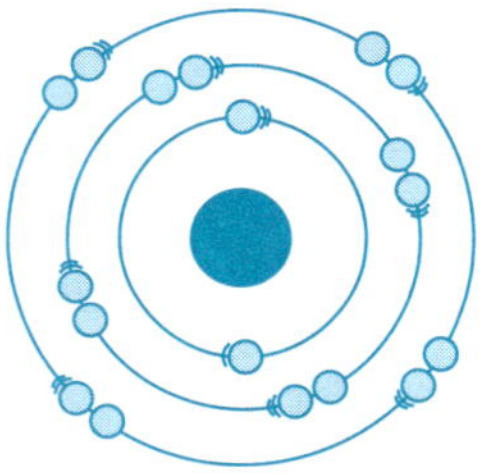

a If this is an atom of argon, are all the electron shells full?

---

b Is this element a solid, a liquid or a gas?

---

c How does it react to electricity being passed through it?

Question 1

In the table below, list the subatomic particles of an atom and their electric charge.

Subatomic particle	Electric charge

Question 2

Complete the following table, which lists the differences between atoms and substances.

Factors differentiating atoms	Factors differentiating substances
1. The type of atom.	1.
2.	2. How many atoms of each element are present.
3.	3.

Question 3

Are the following statements true or false?

a Electrons are particles with a positive electric charge. ______

b It is believed that electrons travel around the nucleus in orbits. ______

c Each atom of an element has a certain number of electrons. This is called the atomic mass. ______

d Atoms of different elements have different size and weight. Chemists call this the atomic weight. ______

e The nucleus contains particles called protons, which have a positive electric charge.

______

f Atoms are electrically neutral. __________

g The number of protons and neutrons added together is called the atomic number of an atom. __________

h Atoms are held together by strong attraction between the protons and electrons. __________

i The nucleus also contains a number of uncharged particles called subatomic particles.

____

j Most of an atom consists of empty space.

____

Question 4

Match the following terms with the correct definition:

compound, element, molecule, atom

a The building blocks of matter.

b Matter made up of one or more pure substances, but of uniform appearance.

c Groups of atoms linked together in a fixed way.

d A substance containing only one kind of atom.

Question 5

A neutral zinc atom contains 30 protons, 30 electrons and 34 neutrons. What happens when the atom changes to the zinc 2+ ion?

Question 6

a What is the name of the special chart into which chemists have arranged all the elements on the Earth?

b Explain the basic format of the chart.

c List reasons why this chart is useful.

Question 7

a What is the atomic number of sodium?

b If the mass number of sodium is 23, how many protons, neutrons and electrons does it have?

c How many electrons are in its outer shell?

Question 8

Provide the correct term for the following statements.

a An atom with an overall positive charge.

b An atom or group of atoms with an electric charge.

c An atom with an overall negative charge.

Question 9

a Are there more metals or non-metals among the elements in the periodic table?

b Provide the properties of the groups of elements in the following table.

Group of elements	Properties
metals	
non-metals	
noble gases	

Question 10

Complete the following information by adding the correct words.

Each element behaves differently from all the others. We say their __________ are different. For example lithium, sodium and potassium have similar properties and are very __________. They are sometimes referred to as the __________ metals. The group containing fluorine also show similarities in how they behave and are called __________.

Question 11

What information can be extracted from the fact that carbon has a mass number of 12 and an atomic number of 6?

____

____

____

____

Question 12

Water (containing both hydrogen and oxygen) has two hydrogen atoms for each oxygen atom. What do you think would be the case with other elements in the same group as oxygen?

____

Question 13

List the ionic charges for the first three elements listed in the relevant columns of the periodic table.

1	2	3	16	17	18

Question 14

Match the following definitions with the correct term.

Compound	Smallest bit of an atom.
Element	A group of atoms.
Subatomic	Pure substance made of particles more than one element.
Molecule	Pure substance made of only one type of atom.

Question 15

Diamonds, buckyballs and graphite are all made of carbon atoms. Explain why these three substances are so different.

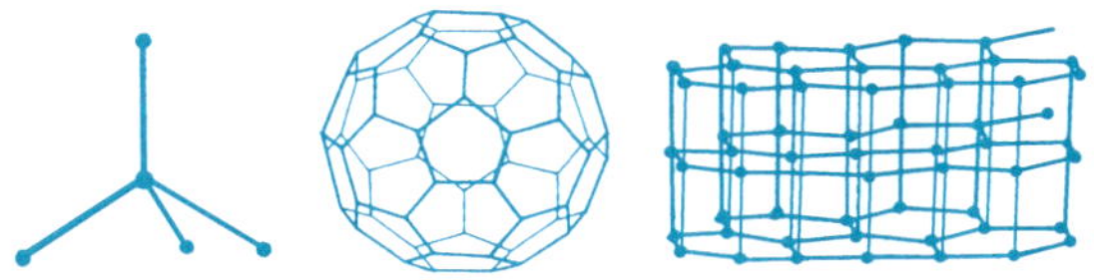

diamond buckyball graphite

____

____

____

____

CHAPTER 4

GEOLOGY AND PLATE TECTONICS

When is the Earth's birthday?

The Earth was formed 4.6 billion years ago. Since then its surface has been scratched, worn, drained, stretched, flattened, flooded and fired. Huge mountain ranges have come and gone and continents of ice have gouged it out. The scars of this abuse can be seen as the flatlands, valleys, mountains and hills that we see every day.

Deep beneath our feet the inside or **mantle** of the planet is churning with awesome power that can make the ground tremble, throw up mountains and volcanoes and force continents to collide or split apart as they ride on the giant conveyor belts of molten rock circulation beneath the Earth's **crust**. The Earth's structure ranges from a twisted, tortured, crumpled and solid broken shell, down to a dense, metallic core where pressures and temperatures are immense.

Continental drift

The world is a moving place. Our planet's rocky surface is a restless jigsaw of about twenty slowly moving pieces. During the 1960s geologists came to realise that the crust, or lithosphere, is broken up into plates that move with the currents of the mantle. The theory of global dynamics, in which the lithosphere moves in response to the **convection** in the upper mantle, is called the theory of **plate tectonics**.

The plates, which carry the continents and seas, have drifted for hundreds of millions of years.

Where the Earth's crust is thick there are continents, where it is thin there are seas.

Tectonic plates

The continents as we know them developed from one giant continent called **Pangaea**. **Pangaea** began to break up about 225–200 million years ago into two super continents, one in the north called **Laurasia** and one in the south called **Gondwana**. These two super continents fragmented into the continents we have today.

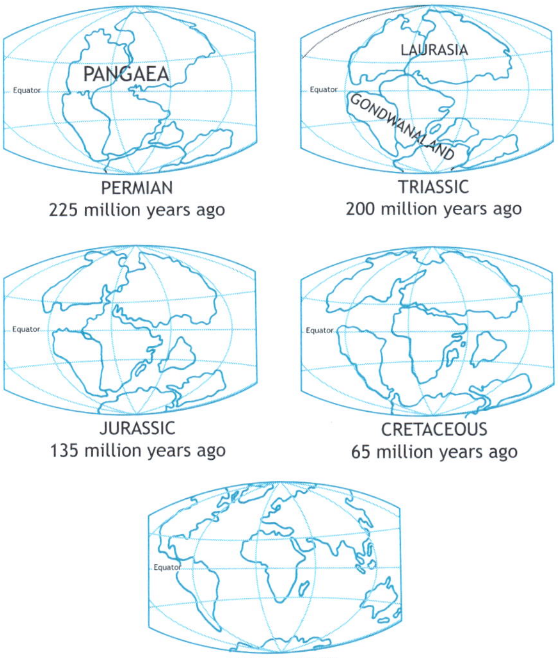

PERMIAN
225 million years ago

TRIASSIC
200 million years ago

JURASSIC
135 million years ago

CRETACEOUS
65 million years ago

PRESENT DAY

The edges of tectonic plates are sites of intense geological activity such as earthquakes, volcanoes and mountain formation. The plates are always moving, sliding over and under each another and even colliding with each other. They move very slowly over small distances of around 2.5 – 5.0 cm a year.

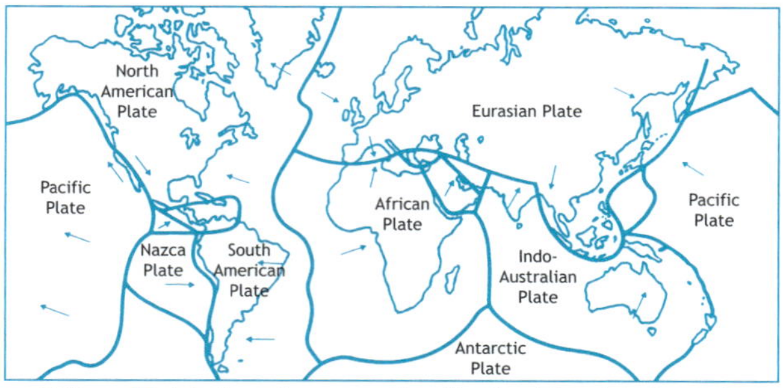

The Earth's tectonic plates

Locations of plate tectonics

The tectonic plates are continually being and destroyed along the mid-Atlantic trenches.

There are three main locations of plate activity:

- **Constructive boundaries** – When two plates move away from each other, they leave a gap out through which magma, from the Earth's mantle, oozes. Huge chains of volcanoes form this way. The mid-Atlantic trench was discovered in the 1950s in the middle of the Atlantic Ocean. It is 60 000 km long and winds through all the oceans of the world. Magma also oozes up from the mantle to build new seabed.

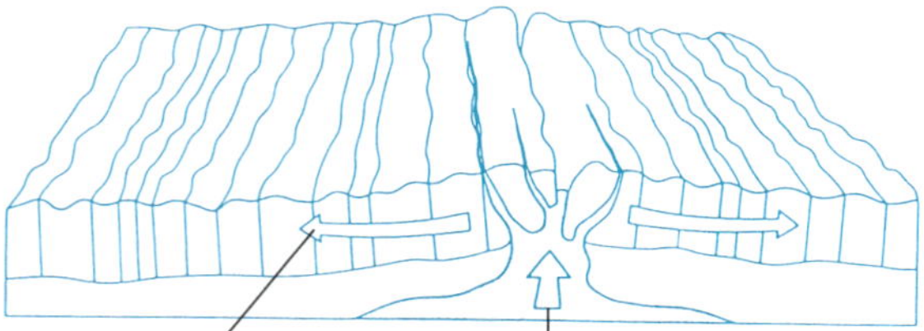

Mid-atlantic trench – constructive plate boundary

- **Destructive boundaries** – Sometimes plates push against each other and one is forced down under the other. The old ocean crust is being dragged back into the mantle. This process is called **subduction**.

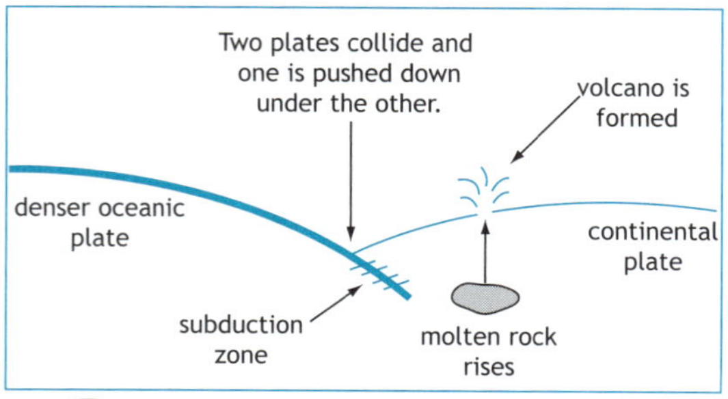

Subduction – destructive plate boundary

- **Intraplate hot spots** – Sometimes a narrow section of magma burns through the crust creating a 'hot spot'. This section remains stationary while the plate moves over it. The Hawaiian Islands and the Great Dividing Range were formed this way.

Geology's superstars

- **Earthquakes**

Earthquakes are associated with destructive plate boundaries. Sometimes, as two plates scrape together, one gets caught on the other. Pressure builds up until it is finally released and felt as an earthquake. Small shocks cause avalanches, landslides and tidal waves. Earthquakes associated with the San Andreas Fault on the West Coast of the United States are often much more serious. Pressure builds up as the Pacific and North American plates grind against each other.

Seismologists measure the strength of an earthquake using a seismograph. The Richter Scale ranges from 1–10 and measures the size of an earthquake. A reading of 7 on the Richter Scale is a fairly strong earthquake. A reading of 1 is not.

- **Tsunamis**

Tsunamis are gigantic destructive waves mostly caused by earthquakes. As the wave approaches shallow water it slows down, gets higher and finally slams into a coastline.

- **Volcanoes**

A volcano is like a safety valve in the Earth through which **magma**, gases, solid rock fragments, poisonous fumes and ash can be discharged. Underneath volcanoes, in the Earth's mantle, are pockets of magma. As heat and pressure build up in these magma chambers the gases expand, which finally causes an eruption.

When the magma reaches the surface of the Earth it is called **lava**. Lava can flow very fast and it cools quickly, forming new igneous rocks such as obsidian, pumice and basalt. These are called **extrusive rocks**. Sometimes the magma does not reach the surface, it forces its way instead into or beneath areas of rock in the Earth's crust. It cools very slowly, forming the **intrusive rock** granite, which forms structures such as laccoliths, dykes, sills and batholiths.

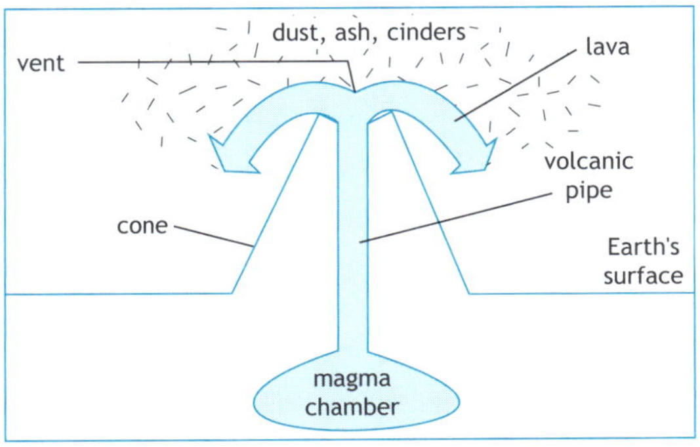

Profile of a volcano

There are about 1000 active or potentially active volcanoes in the world. The greatest activity is in the 'Ring of Fire' around the Pacific Ocean. Australia, however, has very few volcanoes or earthquakes because we do not have plate boundaries crossing the country.

Volcanic landforms

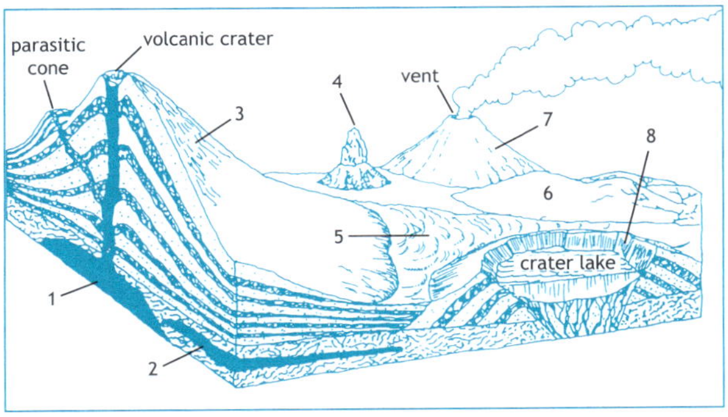

1. Magma chamber, molten rock within the Earth.
2. Lava flow, within rock layers under the ground surface forms intrusive igneous rocks.
3. Composite cone, alternate layers of ash and lava.
4. Volcanic plug, resistant core from an eroded volcanic cone.
5. Lava flow, from the composite cone.
6. Ash deposits, from ash showers.
7. Cinder cone, successive layers of volcanic ash, mud and rock fragments.
8. Caldera, steep-sided crater caused by a violent volcanic explosion.

• Mountain building

Plates can also push against each other, compressing and folding the land upward into mountains over millions of years.

The Himalayan Mountains started to form 25 million years ago and are still forming as the Indian and Asian plates push against each other. Scientists believe that these mountains are increasing in height by 8 cm per year.

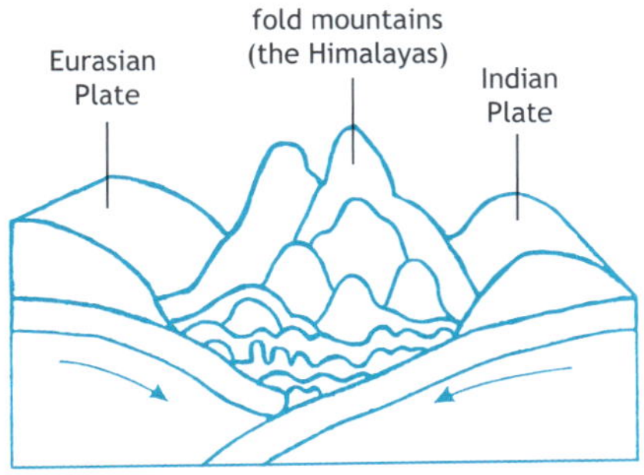

Formation of the Himalayan Mountains

The Earth may also crack under pressure. This is called faulting. A normal fault is when one part of the Earth's surface tilts away from the other. A reverse fault is when one part rides up beside another. The Great Rift Valley of Africa has been formed by faulting. Some examples of surface features caused by faulting are shown in the diagram below.

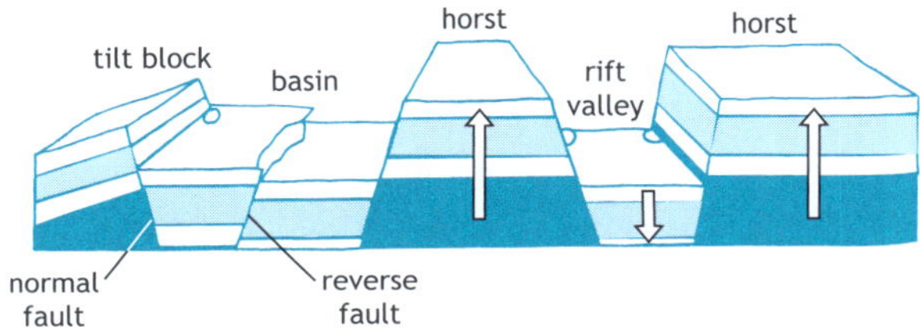

Surface Features

The rock cycle

The rock cycle is the cycling of rock through the Earth, brought first from the mantle as magma to the surface as lava via volcanoes. This cooled rock is then weathered and fragments are eroded and carried to the sea by wind, rain and rivers. This debris collects on the ocean bed, solidifies and is drawn back to the mantle at subduction zones.

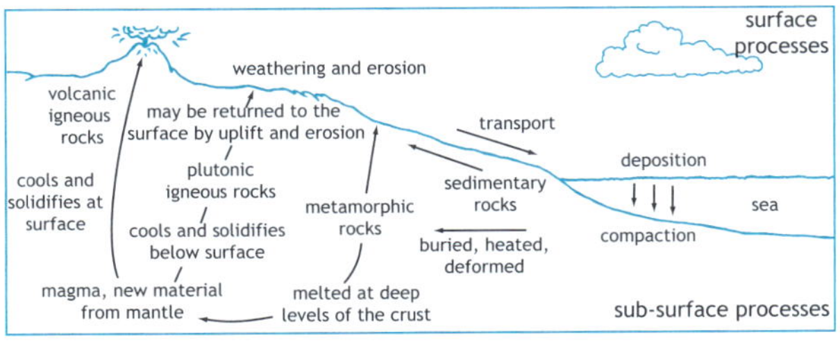

The rock cycle

The Australian continent

The crustal plates carrying the southern continents started to move apart about 80 million years ago. About 65 million years ago Australia and Antarctica were still narrowly connected. The vegetation of Australia and Antarctica was cool, temperate rainforest. About 55 million years ago Australia and Antarctica started to separate. Australia headed north at about 8 cm per year, carrying with it the plants and animals of the Gondwana rainforest. The climate began to change, the sea flowed into the north western areas and the eastern highlands began to form.

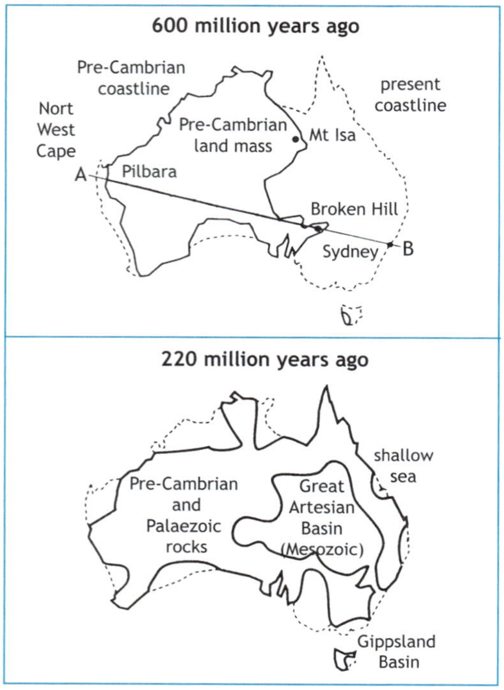

Australia's coastline

Australia has been a separate continent for 45 – 50 million years. This isolation allowed for the evolution of large marsupials, which are now extinct and known only in the fossil record, and many of Australia's current native plants and animals are found nowhere else in the world.

Australia's geology, biology and fossil record provide evidence which supports the theory that all the southern continents were once connected. There are similarities in geology and fossils along the edges of the continents and some plant and animal species are closely related. For example the group of flightless birds called ratites occur in New Guinea (cassowary), Australia (cassowary and emu), New Zealand (kiwi), South America (rhea) and Africa (ostrich). This has now been confirmed through recent DNA comparisons.

Throughout history the continents have drifted to new latitudes, carrying plants and animals that had evolved by that time. The oceans have also been opened up or lost.

Plate tectonics is still an active process and will drastically reshape the face of the Earth over the next 50 million years. For example, scientists are predicting that:

- parts of California will separate from North America
- the Italian 'boot' will disappear
- Australia will become linked to Asia
- Africa will separate from the Middle East.

Glossary

Batholith – A large body of intrusive igneous rock.

Continental drift – The theory that the continents have moved in relation to one another.

Convection – The transference of heat by the movement of the heated parts of a liquid or gas.

Core – The central part of the Earth.

Crust – The outer most layer or shell of the Earth.

Dyke – A vertical igneous intrusion across strata.

Earthquake – A series of waves propagated in the Earth, initiated where stress along a fault exceeds the elastic limit of the rock so that sudden movement occurs along the fault.

Extrusive rock – A type of igneous rock formed when magma erupts onto the Earth's surface and then cools.

Fault – A surface along which a rock body has broken and been displaced.

Fold – A bend or flexure in a rock.

Gondwana – The ancient continental landmass that is thought to have split apart to form the present-day continents of South America, Africa, India, Australia and Antarctica.

Horst – A fault block that has been uplifted in relation to the adjacent rocks.

Intrusive rock – A type of igneous rock formed when hot magma cools beneath the Earth's surface, becoming embedded in older rock layers.

Lithosphere – The relatively rigid outer zone of the Earth.

Laccolith – An igneous intrusion that has arched up the rock strata into which it was forced, forming a lens-shaped body with a flat floor.

Laurasia – The ancient landmass that is thought to have split apart to form Europe, Asia, North America and Greenland.

Mantle – The zone of the Earth's interior between the crust and the core.

Pangaea – A hypothetical continent from which the present continents originated by plate movement.

Plate tectonics – The theory of global dynamics in which the crust is broken up into plates that move over the mantle. The margins of the plates are sites of considerable geologic activity.

Rift valley – A valley formed by a block faulting pulling the crust apart.

Seismologist – A scientist who measures seismic waves or vibrations produced within the Earth.

Sill – A horizontal body of intrusive rock forced between layers of enclosing rock.

Subduction – Subsidence of the leading edge of a lithospheric plate into the mantle.

Tsunami – A seismic sea wave caused by an earthquake, faulting or a landslide on the sea floor.

Volcano – A vent in the Earth's crust from which lava and other materials are ejected.

Science skills

Question 1

The map below shows the major tectonic plates and their present direction of movement.

a Which plate is the African plate pushing against?

b Which mountain range is being pushed up by the Indo-Australian plate?

c What problem is the Pacific plate causing in Japan?

d What is happening along the west coast of South America?

e On which plate is Australia?

f Is there much earth movement in Australia? Why?

g Which two plates are causing the 'Ring of Fire'?

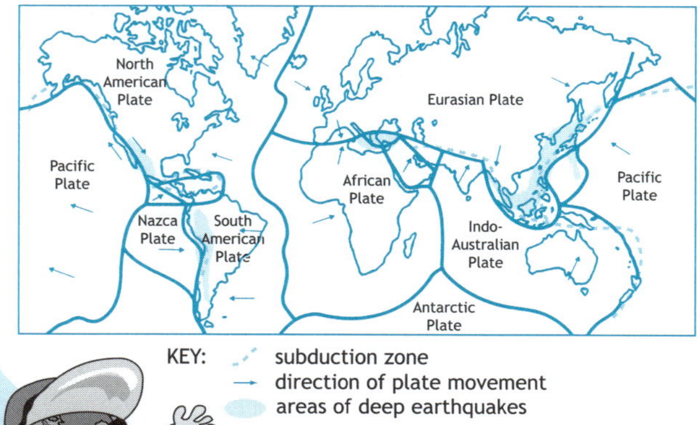

Question 2

Evidence suggests that Australia first appeared in the Pre-Cambrian era. The following diagram is a cross-section of the Australian continent.

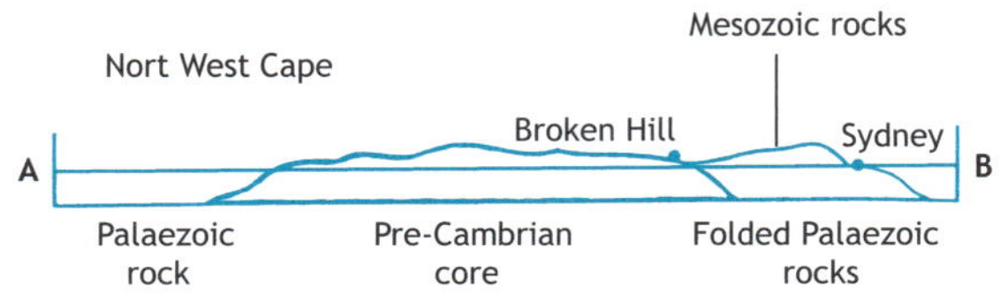

a In which era was the Australian continent's core laid down?

b What happened to this rocky core, being exposed to the weather for millions of years?

c Where did all the eroded material get deposited?

d In which era did this occur?

e What was happening on the eastern seaboard in the Palaeozoic era?

Question 3

Label the diagram of the rock cycle below by putting the correct letters in the boxes provided.

a volcanic igneous rocks

b weathering and erosion

c compaction

d melted at deep levels of the crust

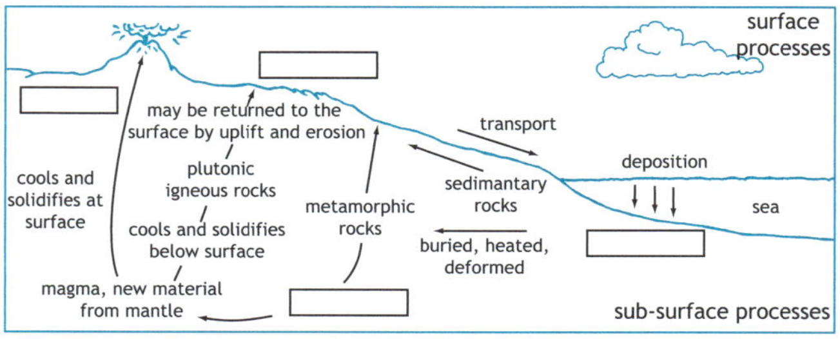

Question 4

This map shows the main areas of fold mountains. What earth movements would also be found in these areas?

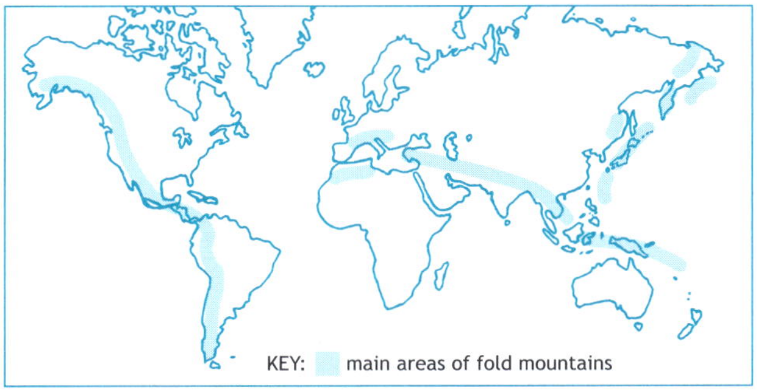

Question 5

A crater lake is formed by a violent volcanic explosion.

Match the diagrams with the descriptions and put them in the correct order.

i Once the gas pressure in the magma chamber drops below a certain point, the volcano can no longer support the cone. The sides slump creating a crater-shaped cavity.

ii As the magma level declines, ash eruptions continue. This further lowers the pressure in the magma chamber.

iii If magma percolates into fractures beneath the collapsed volcano, small volcanoes may form in the crater floor and water may collect.

iv A crater lake starts forming as repeated eruptions begin to deplete the volcano's magma supply. The gas pressure in the magma chamber drops.

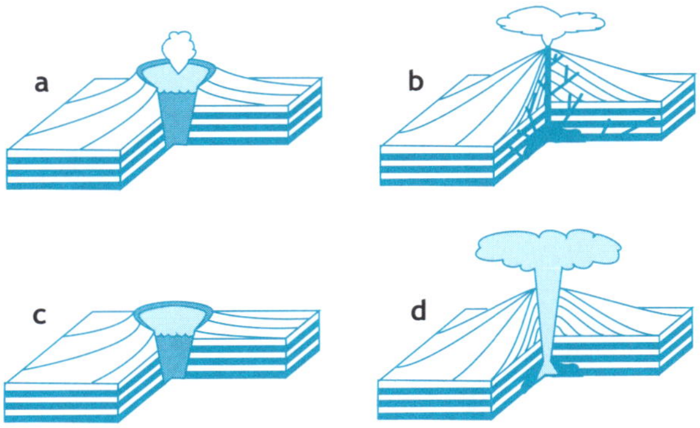

Question 6

Match the following terms with the correct explanation.

Volcano	A common, coarse grained, light coloured intrusive rock.
Magma	A horizontal body of intrusive rock forced between layers of enclosing rock.
Granite	A vent in the Earth's crust from which lava and other materials are ejected.
Lava	Hot molten rock generated within the Earth.
Sill	An igneous intrusion that has arched up the rock strata into which it was forced, forming a lens-shaped body with a flat floor.
Laccolith	Hot molten rock that reaches the surface of the Earth.

Question 7

Label the diagram of a volcano below with the following terms:

cone, volcanic pipe, lava, magma chamber, Earth's surface, (dust, ash and cinders)

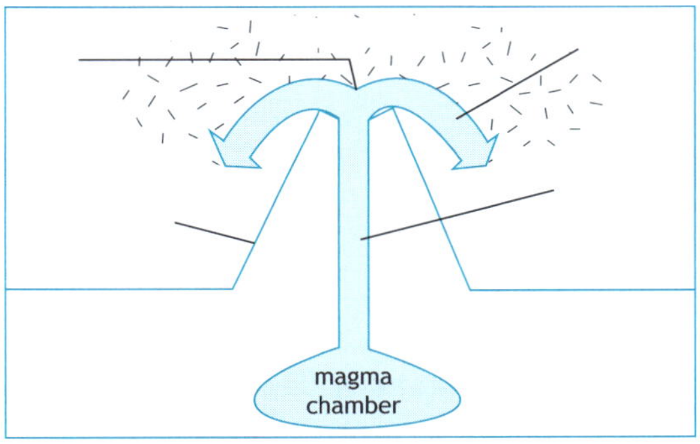

Question 8

Match the correct diagram to the explanation.

i The plates push together and fold mountains are formed as one plate slides under another. Magma can also flow into the folded crust and volcanoes form.

ii The plates slide past each other. The weakness can lead to volcanoes and earthquakes.

iii The plates move apart. Magma flows up into the gap and new rocks are formed.

a

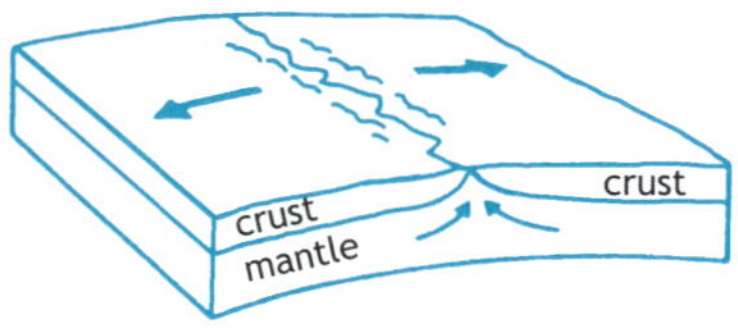

b

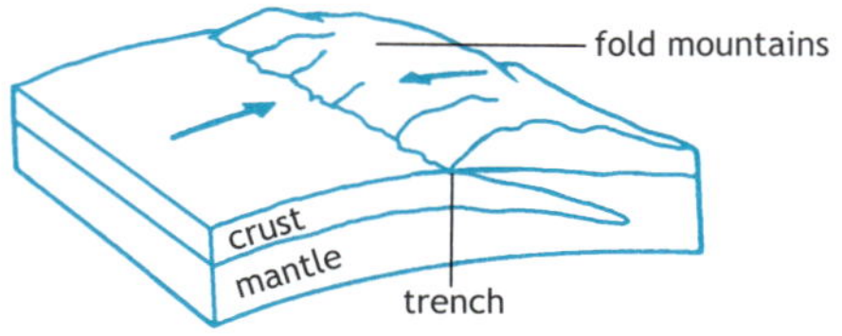

c

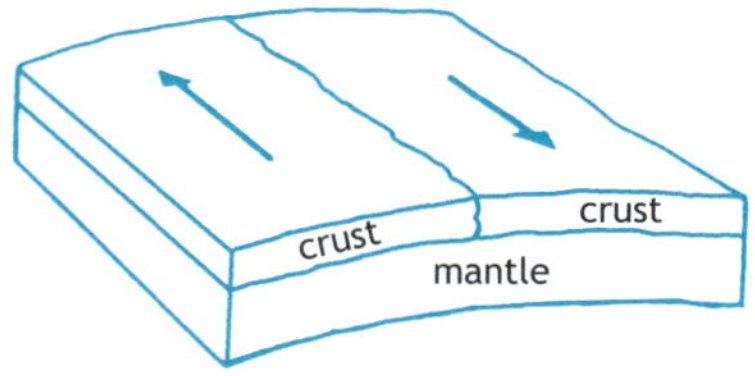

Question 9

Plates are large pieces of rigid crust floating on the mantle. The thickness of a plate varies but they are usually about 100–200 km thick.

Study the diagram below and add the following terms:

crust, continent, convecting mantle, ocean, lithosphere, plate, rigid mantle

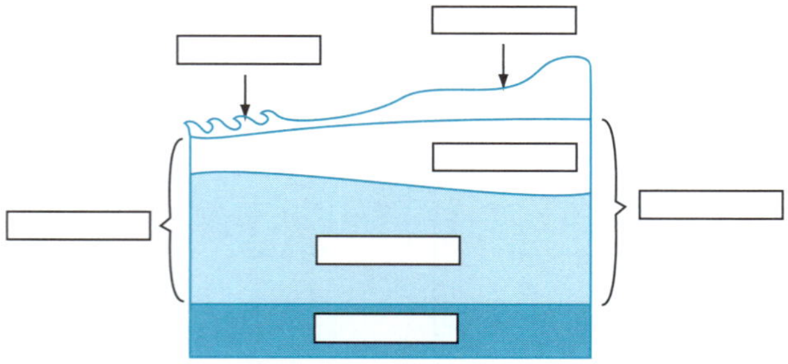

Question 10

Label the following diagram of the internal structure of the Earth with the terms:

core, mantle, crust and lithosphere

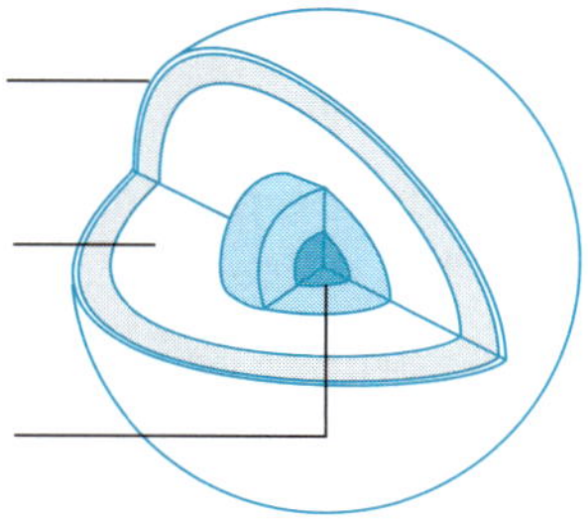

Question 11

The theory of continental drift is supported by a variety of evidence. The following diagram shows several species of flightless birds and the countries in which they are found.

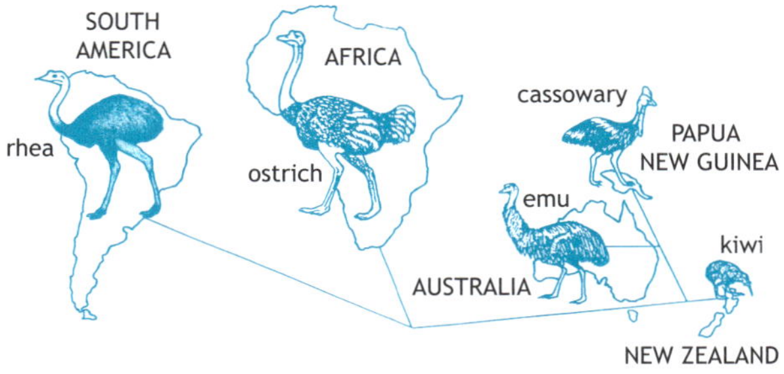

a Which flightless bird is found in South America?

____

b Which is the smallest flightless bird?

____

c Explain how these flightless birds provide evidence of continental drift.

____

____

____

____

d List two other significant pieces of evidence supporting continental drift.

____

____

Question 12

Label the diagram below with the names of the following continental plates:

African plate, Eurasian plate, Indo-Australian plate, Pacific plate, North American plate, South American plate, Nazca plate, Antarctic plate.

Question 13

Below is a table containing some of the world's landforms. Identify which are the result of constructive plate boundaries (C), and which are the result of destructive plate boundaries (D).

Landform	Boundary
The Great Dividing Range	
The Himalayan Mountains	
Andes in South America	
Hawaiian Islands	
Mid-Atlantic Trench	

Question 14

Label the diagram below with the following terms:

normal fault, reverse fault, horst, rift valley, tilt block

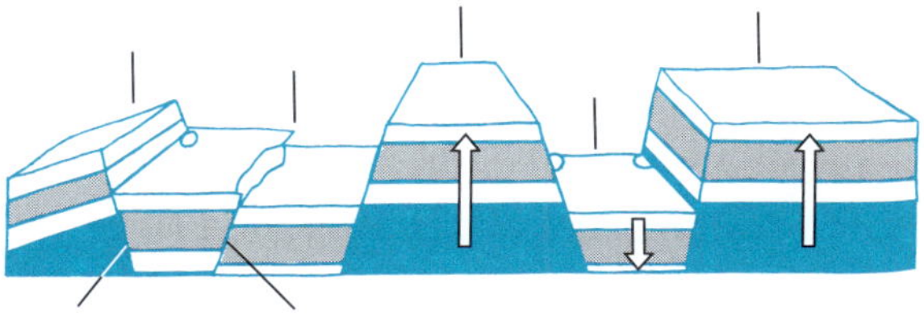

Question 15

The following diagrams show the movement of the Australian continent over the past 135 million years.

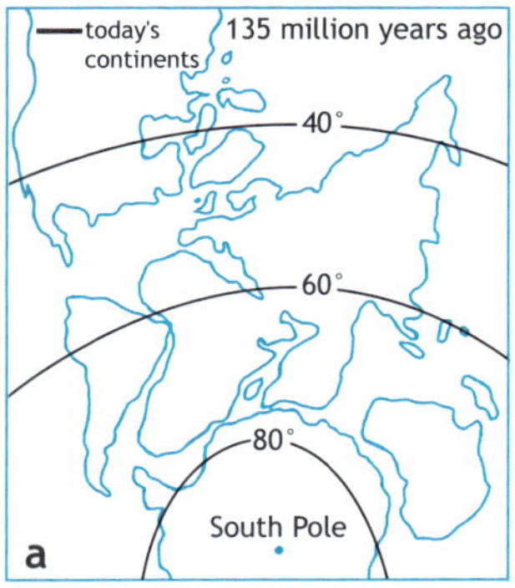

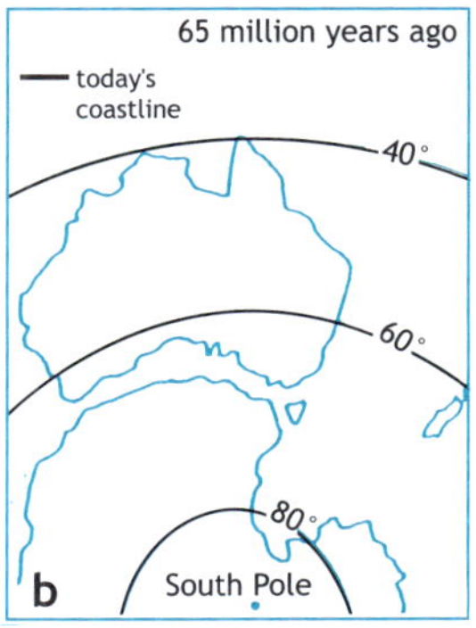

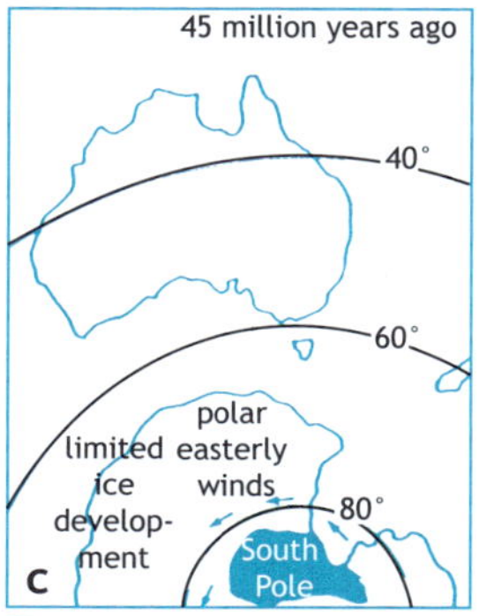

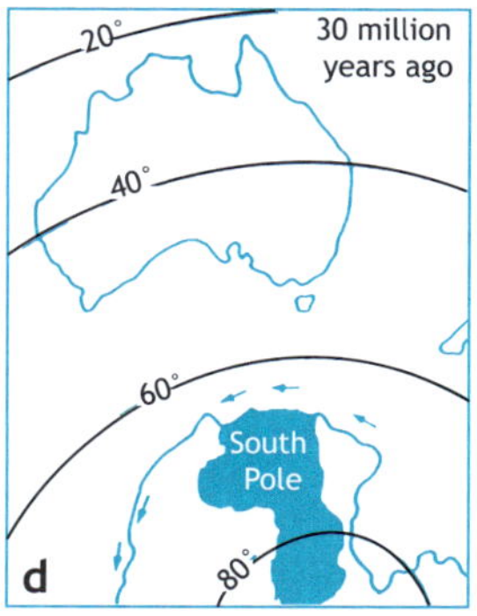

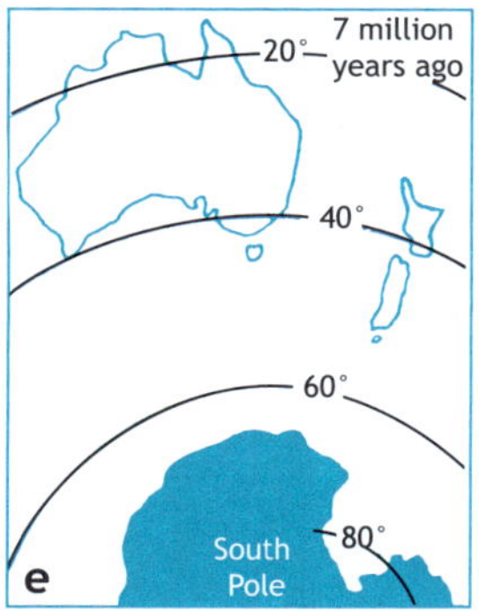

a Between which latitudes did Australia lie 135 million years ago?

b How long did it take for Australia to move completely north of the 60 degree latitude?

c How many degrees south was Sydney 65 million years ago?

d Approximately which latitude is Australia on now?

e What has happened to Australia's climate over the last 135 million years?

Revision questions

Question 1

The plates of the Earth's crust float on the underlying mantle. The following diagram shows the boundaries of two sets of plates.

a Name both plate interactions.

b Explain what is happening in each diagram.

c What is the name of the process occurring in diagram B?

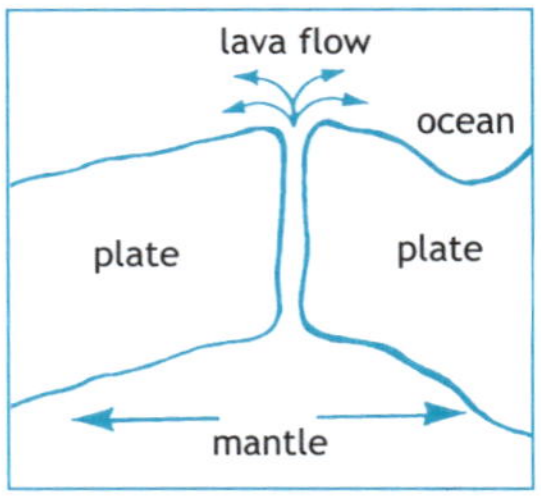

A

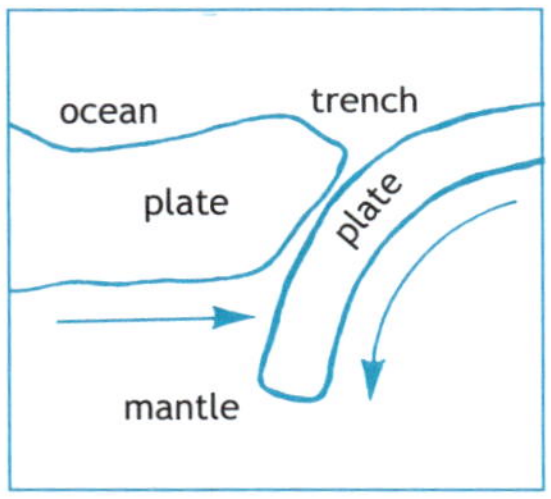

B

Question 2

a What causes volcanoes to erupt?

b Where do volcanoes occur?

c What is the name of the ring of volcano activity surrounding the Pacific plate?

d What is the main type of rock formed if magma cools above ground?

e What is a dyke?

Question 3

Complete the sentences below.

A ________ is a gigantic destructive wave caused mainly by underwater ________. As it approaches ________ water it slows down and gets ________. Finally the wall of water slams into a coast.

Question 4

Marsupials are abundant in Australia and South America and they are also common in New Guinea. Fossils of mammals have been found in Antarctica but there is absolutely no evidence of marsupials in Asia. Scientists believe that Australia and South America were both part of Gondwana at the same time, but India, Africa and New Zealand had split off before marsupials developed.

a Which theory does this evidence support?

b How could you explain the existence of fossils in Antarctica?

Question 5

Since the Earth's crust was formed, the continents have always been in motion. Scientists have only been able to track the continent's movement from about 600 million years ago.

a What was the name of the giant continent that existed between 400 and 200 million years ago?

b What are the names of the two super continents that appeared when the giant continent broke up about 200 million years ago?

c Arrange the diagrams below in the correct order of development, from the oldest to present day.

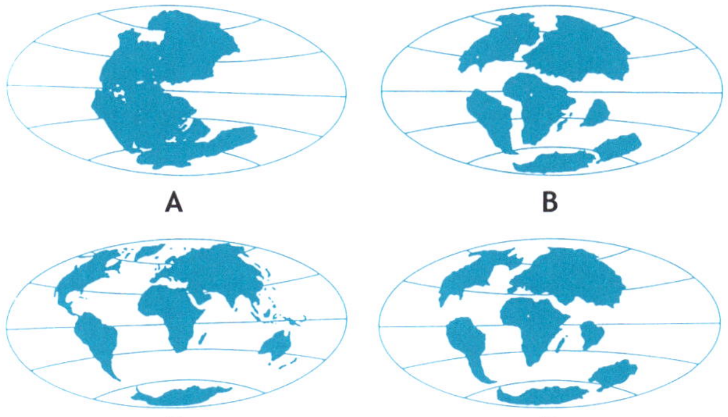

Question 6

The following diagram shows the geological process that formed the Hawaiian Islands.

a Explain what is happening here.

b What is this called?

c What other landform was created this way?

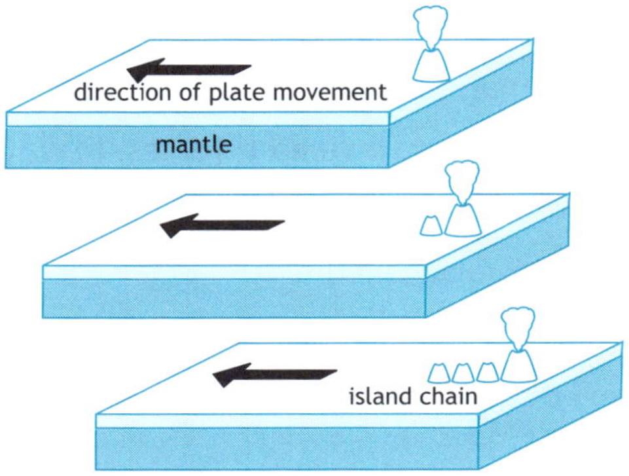

Question 7

Study the diagram below and write in the names of the structures A to D.

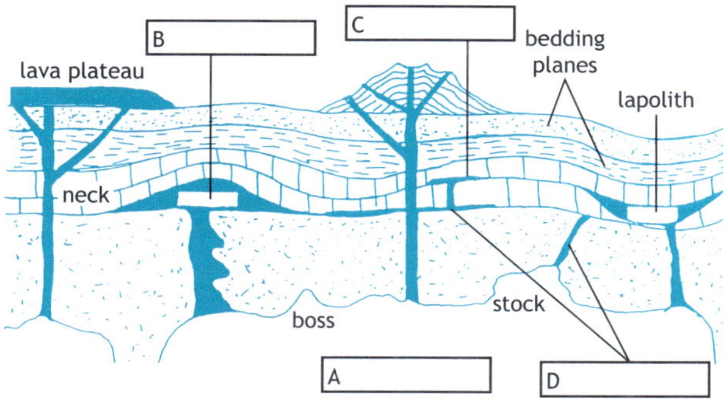

Question 8

Write in the correct words to complete the information below.

It is believed that the Earth's surface is divided up into ________ called ____________ plates, on which the ____________ have drifted about. This continental drift causes ____________ and volcanoes and builds ____________ ranges. These plates are continually being renewed along the ____________ ____________, and ____________ by subduction.

Question 9

Study the diagram below showing three strata of sedimentary rock.

a Which rock is the oldest?

b Which rock is the youngest?

c What is the name of the feature marked 'X'?

d What other change has taken place since the rocks were deposited?

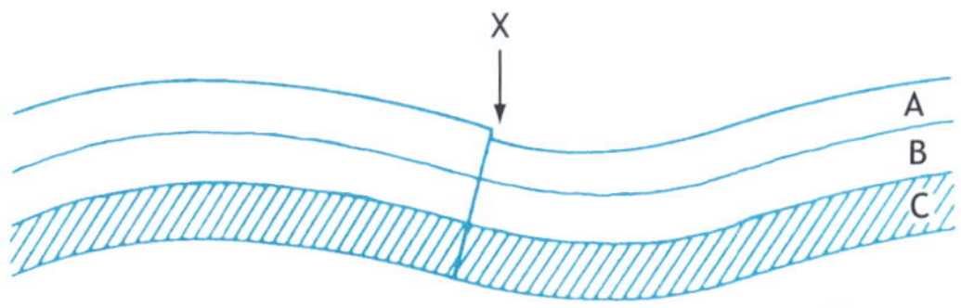

Question 10

The diagram below shows a cross-section through rocks.

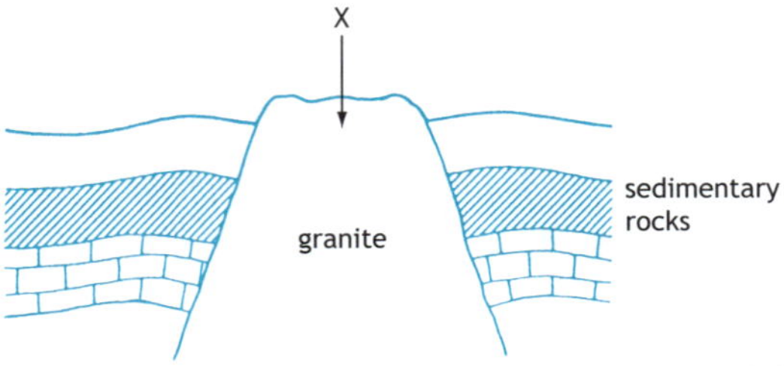

a What type of rock is granite?

____

b How was the granite formed?

____

____

c Why are the layers of sedimentary rock not straight?

____

d How was structure 'X' formed?

____

____

e Why is this granite now exposed to the atmosphere?

____

____

Question 11

Which regions of the world were originally part of the same super continent as Australia?

____

____

Question 12

a What is land called if it is pushed up as a result of faulting?

____

b What do we call land which drops as a result of faulting?

____

Question 13

a Which country is on the same plate as Australia?

____

b What process is forming the Himalayan Mountains?

____

c Name a well-known landform in America that has resulted from a fault.

____

____

d Name a well-known landform in Africa which has resulted from a fault.

____

e List three other landforms produced by volcanic activity.

____

____

____

Question 14

Australia's Pre-Cambrian landmass was very different from the present day formation.

a Which states are missing altogether?

____

b Which basin existed over much of the eastern states about 200 million years ago?

____

Question 15

Fill in the missing information to complete the sentences below.

__________ rocks such as __________, __________ and basalt, result from __________ cooled __________ on the surface of the Earth. Sometimes magma does not reach the __________ of the Earth's crust. Instead it forces its way between rock layers where it solidifies __________ into a coarse rock called __________.

Test A

Question 1

It is known that in a particular area, a sequence of strata of sedimentary rock was laid down on top of one another and has remained in the same order. However, some places in this area have eroded more than others. The four symbols (*, X, O, #) represent four index fossils.

a Explain why we know beds 1 and 8 are the same geological age.

b Indicate two other beds that are the same age.

c Which bed contains the most recent fossil?

d Which is the oldest type of fossil in the area?

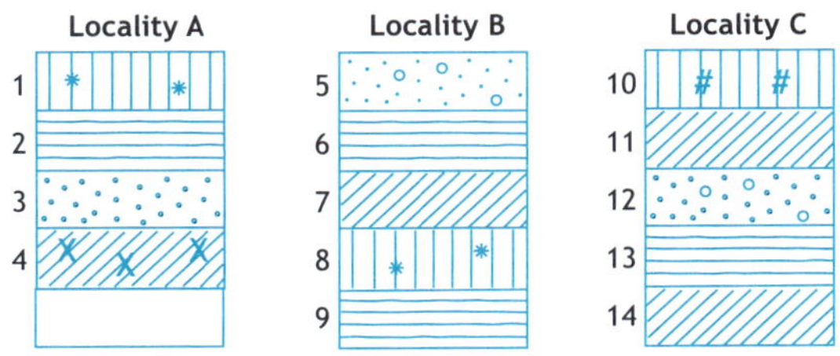

Question 2

The following diagram shows the fossil record recorded in layers of rock. Which of the following statements is supported by the information given?

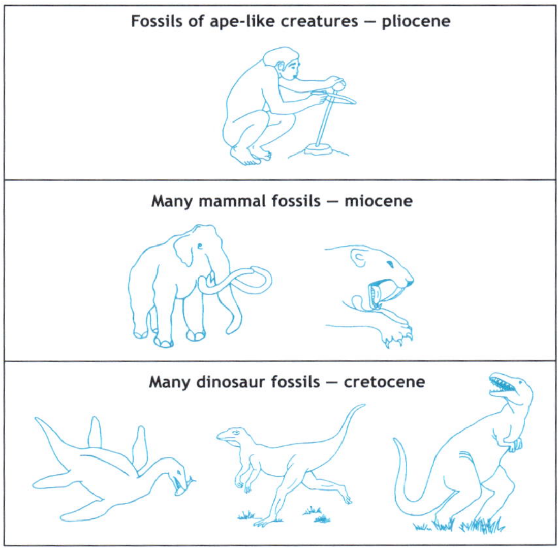

a Ape-like creatures existed before mammals.

b Mammals existed before dinosaurs.

c Dinosaurs existed before mammals.

d Dinosaurs existed before ape-like creatures.

Question 3

During the Earth's history large numbers of species have become extinct in short periods of time.

a What do we call these events?

b How many major events like this do scientists think the Earth has experienced?

c When was the largest of these events?

Question 4

Charles Darwin's theory of evolution is based on three observations:

1. ______________________

2. ______________________

3. ______________________

Question 5

Most environments undergo cycles of change, driven by internal and external forces.

Study the graph below of acacia tree and elephant populations surrounding a river in Africa. The data extends from 1896 until 1990. This river was full in the late 1800s and dried up around the turn of the century. It remained dry until the mid-1950s. In 1982 it dried up again and remains so. Elephants browse on the acacia, but when the river is dry the elephants move to a new area to find water.

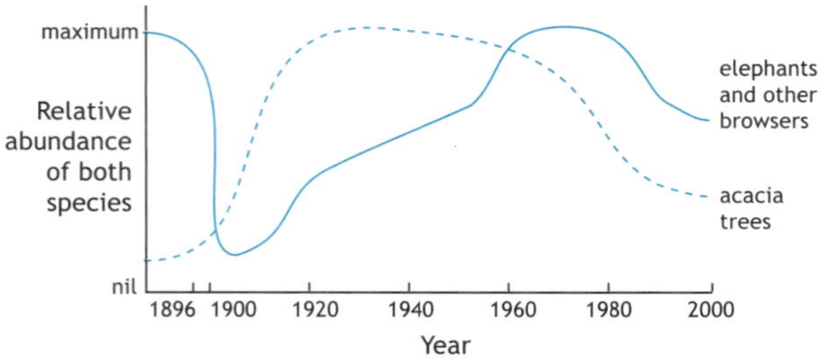

a Mark with arrows the points on the graph at which the river filled and dried up.

b Mark the point when the elephants left the river.

c Why did the population of acacias reach its peak between 1910 and 1950?

____

____

d Why did acacia levels fall after 1960?

____

____

Question 6

Draw a diagram showing and labelling the three basic chemical components of a DNA molecule and how they fit together.

Question 7

Complete the table below with the correct information.

Nucleotide	Full name	Where in the cell	Function
DNA			
RNA			

Question 8

Put the following sentences in the correct order so that they describe the process of protein synthesis.

a Join each new amino acid to the previous one and lift the growing protein off the ribosome.

b Move the messenger RNA out of the nucleus and onto a ribosome where the protein can be assembled.

c Repeat this process for any amino acid and when it is finished the completed protein falls off the ribosome.

d Add transfer RNA molecules to the matching triplets on the messenger RNA.

e Unravel the DNA and copy the code from one strand into the messenger RNA.

Question 9

The two mice below were mated.

BB (male) bb (female)

a What is the genotype of the male?

____

b What will be the genotype of all of the first generation offspring?

____

c What will be the phenotype of all of the first generation offspring?

____

d What will be the ratio in the second generation offspring if the first generation offspring are mated?

____

Question 10

Like haemophilia, colour blindness is a sex-linked condition, carried on the sex chromosomes. It is due to a recessive gene on the X chromosome. If a colourblind male (X^CY) married a female with on recessive gene for colour blindness (XX^C), what percentage of their offspring would be:

a female with at least one recessive gene?

b male, suffering from colour blindness?

Question 11

The cattle below are both homozygous for their horn length. The allele for long horns is the dominant allele. The offspring of these cattle were mated.

a What would be the expected horn length ratio in the offspring of the second generation?

b What are the possible allele combinations for the offspring?

Question 12

The following drawing is of two different sets of chromosomes. Two sets of genes are marked on each. Fill in the blanks with either 'homozygous' or 'heterozygous'.

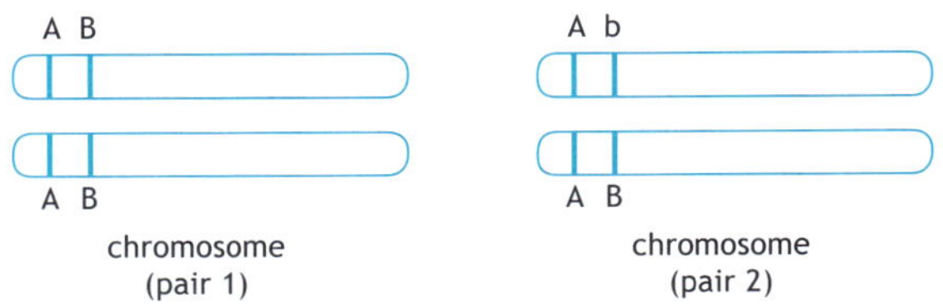

a Chromosome pair (1) are ______________ for gene A and ______________ for gene B.

b Chromosome pair (2) are ______________ for gene A and ______________ for gene B.

Question 13

Which of the following chromosomes display linked genes?

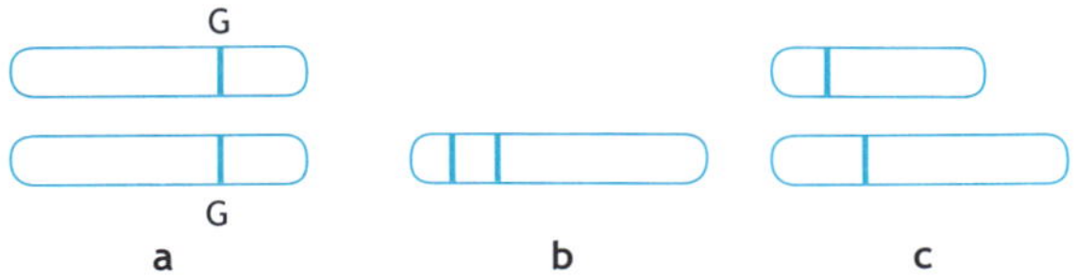

Question 14

The following diagram shows two generations of plants resulting from the crossing of different homologous breeding parents.

a Label the diagram with the terms – parents, first generation, second generation.

b Write the genotypes under each individual.

c Write the phenotypes for the parents, the first and second generations.

Question 15

It has been suggested that a particular characteristic is controlled by a recessive gene located on the X chromosome. In the diagram below the shaded individuals show this characteristic, males are represented by squares and females by circles.

Do you agree with the above suggestion?

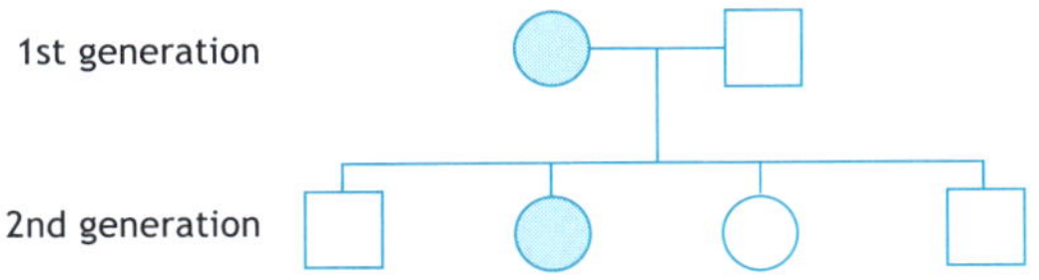

Question 16

Complete the following table.

Atomic number	Mass number	No. of protons	No. of neutrons	No. of electrons
14			14	
		11	12	
	39			19
	14		7	

Question 17

Label the periodic table with the periods and groups.

1 H																	2 He
3 Li	4 Be											5 B	6 C	7 N	8 O	9 F	10 Ne
11 Na	12 Mg											13 Al	14 Si	15 P	16 S	17 Cl	18 Ar
19 K	20 Ca	21 Sc	22 Ti	23 V	24 Cr	25 Mn	26 Fe	27 Co	28 Ni	29 Cu	30 Zn	31 Ga	32 Ge	33 As	34 Se	35 Br	36 Kr
37 Rb	38 Sr	39 Y	40 Zr	41 Nb	42 Mo	43 Tc	44 Ru	45 Rh	46 Pd	47 Ag	48 Cd	49 In	50 Sn	51 Sb	52 Te	53 I	54 Xe
55 Cs	56 Ba	57 La	72 Hf	73 Ta	74 W	75 Rc	76 Os	77 Ir	78 Pt	79 Au	80 Hg	81 Tl	82 Pb	83 Bi	84 Po	85 At	86 Rn
87 Fr	88 Ra	89 Ac	104 Rf	105 Db	106 Sg	107 Bh	108 Hs	109 Mt	110 Ds	111 Rg	112 Cn		114 Fl		116 Lv		

Question 18

Use the periodic table to complete the following table.

Element	Symbol	Atomic number	No. of protons	Atomic mass
lithium				
			6	
		17		
	Ba			
gold				

Question 19

a What is the symbol for platinum?

b What is the name for the element with the symbol Hf?

c What is the atomic number of gold?

d What is the atomic mass of iodine?

Question 20

What is the meaning of the following chemical formula for sucrose?

$$C_{12}H_{22}O_{11}$$

Question 21

Complete the following information by filling in the correct words.

Many metals react with ________ to produce a gas called ____________. They also produce compounds called ________. The compound ____________ ____________ would be produced if magnesium and hydrochloric acid reacted.

Question 22

Each of the following atoms is holding a card showing the number of electrons in its outer shell.

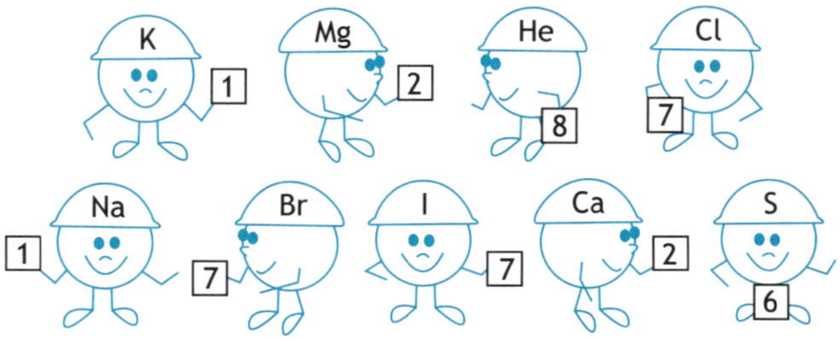

Which metal atoms do you think might react with which non-metal atoms?

____

____

____

Question 23

Using the key below match the following ions with the names provided.

Names: the ammonium ion, the hydrogen ion, the nitrate ion, the hydroxide ion

KEY

Symbol	Element	Atom
H	hydrogen	
O	oxygen	
N	nitrogen	

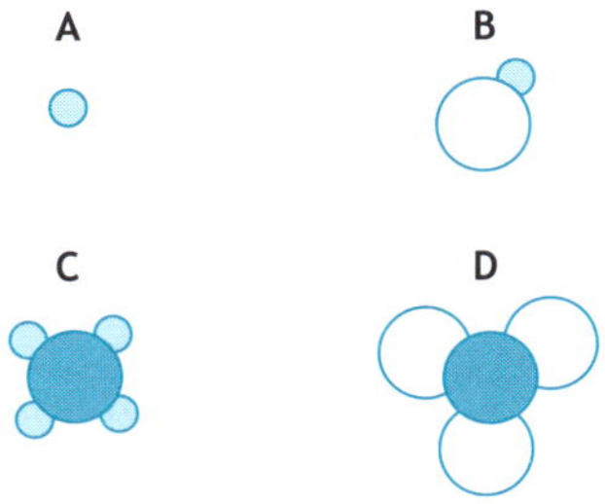

Question 24

Which diagram below best shows this chemical reaction?

$HCl + NaOH \rightarrow NaCl + H_2O$

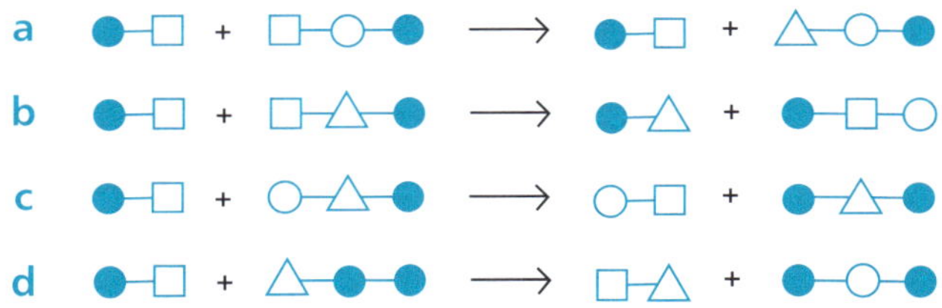

Question 25

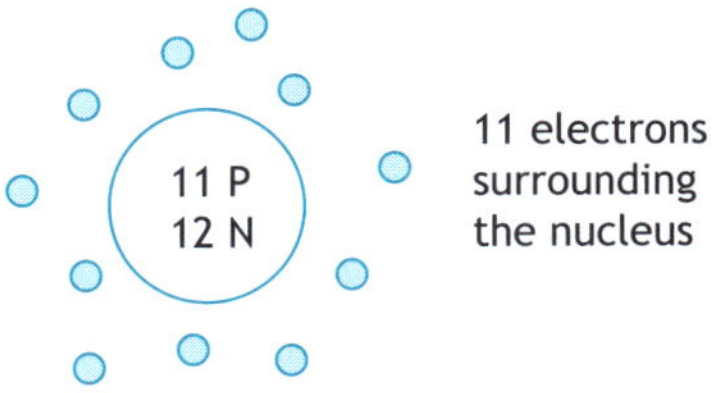

a What is the name of this element?

____

b What charge does it have before the electron is removed?

____

c What charge does it have after the electron is removed?

____

Question 26

Study the diagram of the Earth's structure below and write in the two missing labels.

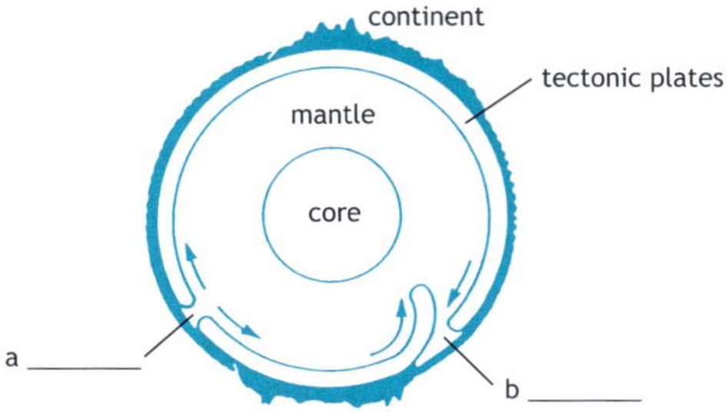

Question 27

Complete the information given below by filling in the correct words.

The Earth's solid ____________ is broken up into several semi-rigid ____________ that move very ____________ when currents in the ____________ come into contact with the underside of the plates. When the plates move ____________, new material wells up and solidifies. This happens at ____________ ____________ beneath the sea and is known as sea-floor spreading. The opposite process, known as ____________, occurs when the edge of one plate ____________ down ____________ another and into the ____________. Because of these two processes, the Earth's ____________ is continually recycled.

Question 28

The diagram below is a simple rock cycle. Number the steps correctly. Start with the box numbered 1.

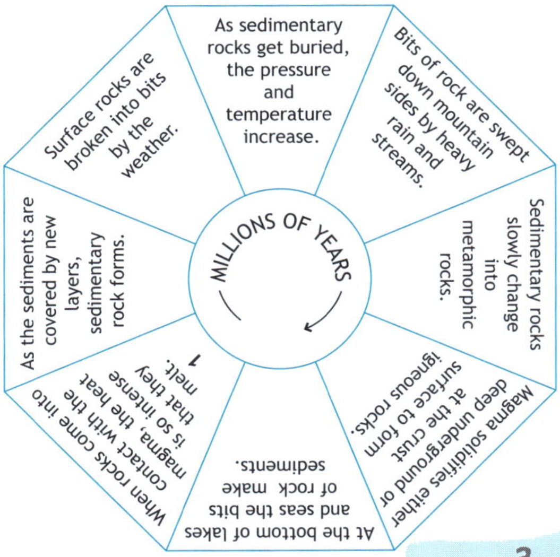

Question 29

Match the following terms with the correct description.

Earthquake	An instrument that measures the amount of ground movement produced by an earthquake.
Basalt	Earth's solid outermost layer.
Crust	A sudden movement of plates in the Earth's crust, caused by the release of strain that has built up over time.
Hot spot	A common fine-grained extrusive rock.
Seismograph	A location where a stationary plume of magma rising from the mantle melts a hole in the Earth's crust, producing a volcano.

Question 30

The diagram below shows a volcano that is no longer active. Mark the following on the diagram:

a The place where you are likely to find volcanic ash.

b The place where basalt is likely to be found.

c The material that is most likely to get eroded away.

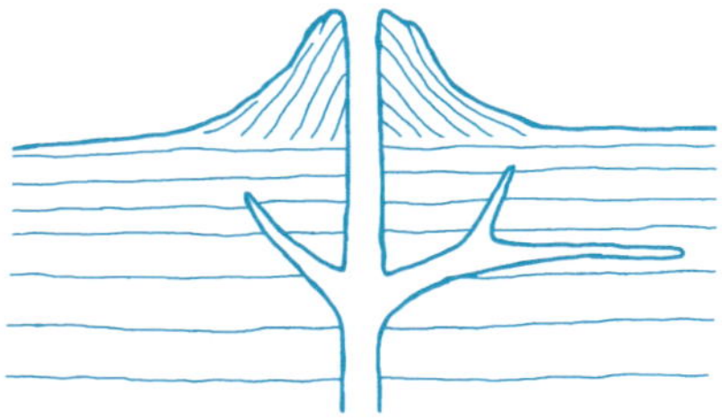

Question 31

The diagram below shows the process of subduction.

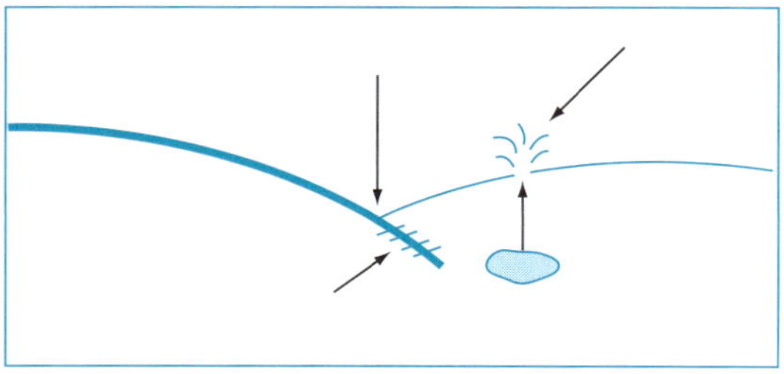

a Explain this process.

b Label the diagram with the following phrases:

subduction zone, molten rock rises, continental plates, volcano is formed, the point where two plates collide

c Which two natural events often occur where subduction is occurring?

Question 32

a What is the name of the scale that measures the strength of earthquakes?

The strength of several earthquakes was recorded in the following table.

Earthquake	Reading
A	4.2
B	1.0
C	7.3
D	5.4
E	8.1

b Which earthquake was the strongest? ______

c Which earthquake probably did very little damage? ______

d Which earthquakes were measured at least 5 on the scale? ______

Question 33

Match the name of the volcanic landform with the correct description.

Composite cone	Within rock layers under the ground surface it forms intrusive igneous rock.
Caldera	The resistant core from an eroded volcanic cone.
Lava flow	A cone formed from alternate layers of ash and lava.
Volcanic plug	Steep-sided crater caused by a violent volcanic eruption.

CHAPTER 5

WEATHER

What if scientists learned to control the weather?

We live at the bottom of a restless ocean of air called the atmosphere. Most of the gases making up the atmosphere occur close to the Earth – this is the region of 'weather'. Some of the gases in the atmosphere are nitrogen, oxygen, carbon dioxide, argon, hydrogen, ozone and water vapour. Without water vapour there would be no major weather systems and no life on Earth.

Meteorology is the study of the atmosphere, its composition, and the changes that take place within it. **Weather** is the term used to describe atmospheric conditions in a specific place at a specific time. **Climate** refers to the average weather conditions of a particular place. These weather conditions are measured and recorded over all the days of the year. Weather records are useful for such things as working out if there is enough solar energy in a particular area for a solar power station, or determining if farming is practical in a certain area. They are also used to predict the weather.

Weather features

Here on Earth the Sun is the main influence on our weather. The Sun's energy stirs the atmosphere into motion, creating wind, rain, clouds and snow. However, the Sun's heat is distributed unevenly around the world, which sets up great currents in the atmosphere and the oceans. These imbalances have a major influence on our weather.

Air pressure

Gravity pulls the atmospheric gases towards the Earth and so they have weight. The weight of the atmospheric gases is called atmospheric pressure. It is strongest at sea level and decreases as the distance from the Earth increases. The international unit for measuring pressure is the **hectopascal**. Air pressure varies from place to place depending largely on the temperature. For example, if the air is cold, the gas molecules move close together and the air weighs more. This is called a high pressure area or a **high**. A low pressure area, or **low**, is when the molecules are warmer and move further apart. High pressure systems are associated with fine weather and light winds, while low pressure systems usually indicate cloudy, rainy weather.

A weather map shows the areas where the air has considerable weight (highs) and relatively little weight (lows). Lines called **isobars** connect areas of the same air pressure and are drawn around the high and low pressure areas. If the numbers get lower moving inwards it is a low, if they get larger it is a high.

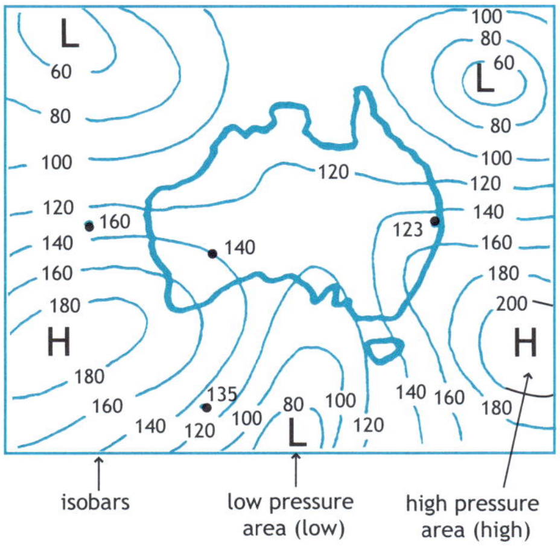

Highs and lows on a weather map

The highs and lows move from left to right across Australia.

Wind

When air begins to move it becomes wind. When a pocket of air is warmer and lighter than the surrounding air, it tends to rise. When it is colder and heavier it tends to sink. Air also moves from high pressure areas to low pressure areas. When isobars are close together there is large difference in pressure, which causes strong winds. If isobars are further apart the winds are gentle. When you see tightly packed isobars around a low, conditions will be very windy.

Winds do not blow in straight lines; they are deflected by the rotation of the Earth – to the right in the northern hemisphere and to the left in the southern hemisphere. This is called the **Coriolis Effect**. The winds mostly follow the isobar lines and in the southern hemisphere move clockwise around a low and anticlockwise around a high.

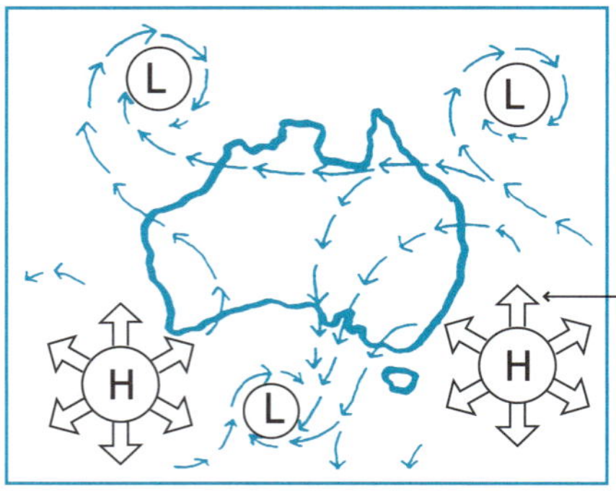

Rotation of the Earth bends the wind to the left – the Coriolis Effect

Wind brings changes in weather. For example, if winds blow over water they are moist; if they blow over the inland they are dry; if they come from the south they are cold and if from the tropics they are warm.

Humidity

Air contains water vapour. The amount of water vapour in the air at any one time is called the **humidity**. The warmer the air is the more water vapour it can hold and the higher the humidity. **Relative humidity** is measured using a hygrometer, which consists of two thermometers – one kept dry and one wet. The smaller the difference between them the higher the humidity.

Clouds

Two things are needed for clouds to form:

- tiny particles (dust, soot, pollen)
- warm, moist air that is cooled.

Colder air cannot hold as much water vapour as warm air, so as moist air rises and cools, water condenses to form water droplets. For clouds to form, the water vapour must have something to condense around. Billions of particles with tiny water droplets around them form a cloud.

There are many different types of clouds, which are classified by the height of their bases above the Earth. The three main types of cloud are cirrus, stratus and cumulus.

Rain

Water in nature is always moving – rising up to form clouds and falling down again as rain. **Precipitation**, or rain, occurs when the droplets in clouds combine and become so heavy that air currents can no longer hold them up. Clouds may also reach up into the colder, higher parts of the atmosphere and the water droplets will turn to ice (hail) or snow.

Forecasting the weather

Weather forecasts are made based on data recorded daily using standard instruments and supplemented with satellite information.

- **Temperature** is a measure of the degree of heat in the air. The maximum and minimum temperatures are measured in degrees Celsius, using a thermometer.

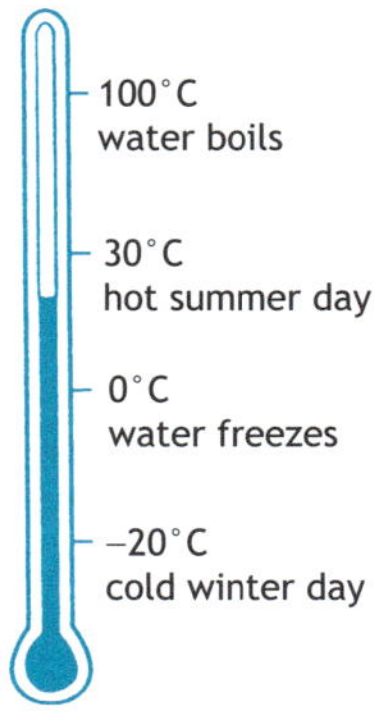

Thermometer

- **Wind speed** can be measured using an instrument called an anemometer. It can also be estimated using the Beaufort Wind Scale (see page 67), which is based on the effect the wind has on the surroundings.

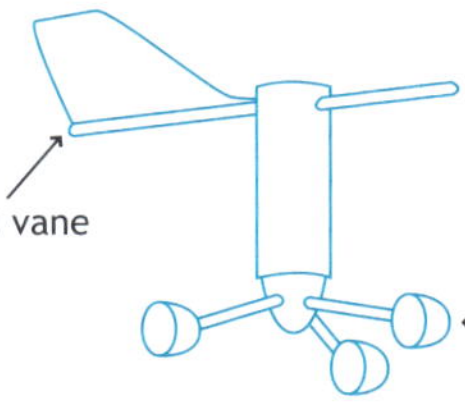

Anemometer

- **Wind direction** is very important in estimating the type of weather change brought by the wind. It is measured using a wind vane or a wind sock. The tail on the wind vane or the mouth of the wind sock points into the wind and indicates the direction from which the wind is blowing.

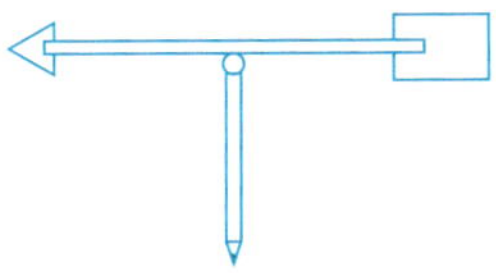

Wind vane

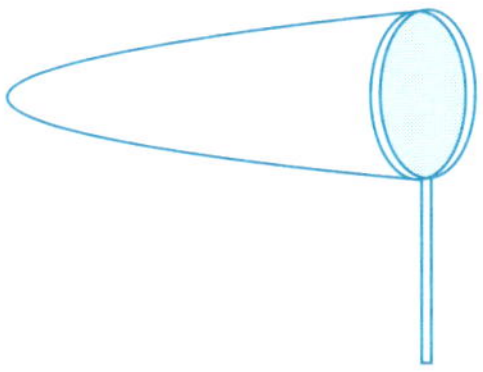

Wind sock

- **Atmospheric pressure**, or the weight of the air at a specific location, is recorded using an instrument called a barometer. A falling barometric reading indicates strong winds and storms on the way.
- **Rainfall** is measured in a rain gauge. This instrument consists of a can, a funnel and a graduated measure. Rainfall is measured in millimetres. On a weather map, areas where rain has fallen are shaded.

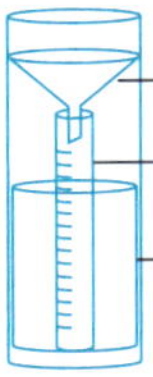

Rain gauge

Humans cannot control the weather, but we do try to predict it. No forecast is ever perfect, as there are many things that meteorologists don't yet understand.

Extreme weather

Weather-related disasters, including floods and droughts, result in enormous damage to crops, buildings, transport and communications, and can cause great social and economic hardship.

- **Thunderstorms** – A thunderstorm is a storm accompanied by heavy rain, thunder, squalls, gusty winds and lightning. They don't last very long but can cause severe flooding. The most spectacular thunderstorms occur in the humid tropics.
- **Cyclones** – A cyclone is a closed circulation of air around a centre of low pressure. In the southern hemisphere, cyclones rotate clockwise. Cyclones are associated with severe storms and very strong winds (gusting up to 200 km/h). They can be up to 400 km across and are very destructive.

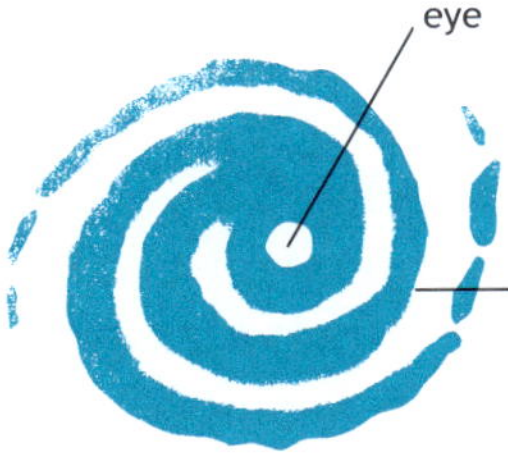

Cross-section of a cyclone showing the eye (low) and the spiral cloud bands

- **Hurricanes** – A hurricane is a spiralling low pressure system formed over tropical seas. This is the most violent of all storms. It is a massive spiral of cloud up to 1000 km in diameter, with driving rain and devastating winds of up to 300 km per hour.
- **Tornadoes** – A tornado is a violently rotating column of air up to 100 m across. The air spins around a central core at speeds of up to 600 km/h. These winds are able to uproot trees, demolish houses and even fling cars into the air.

A tornado

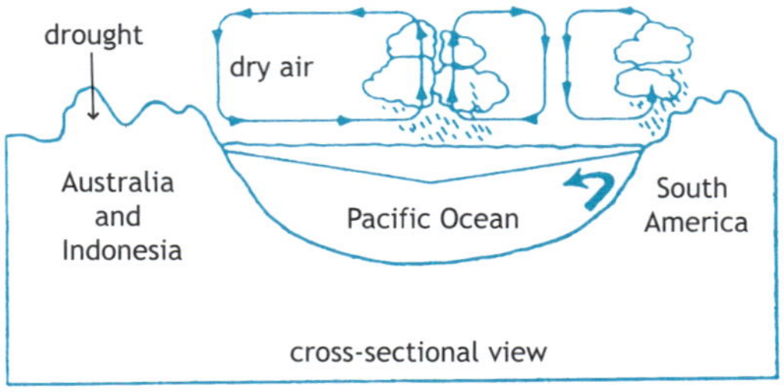

The ENSO situation

Australia's climate and El Nino

Australia not only has one of the most variable climates in the world, but it is the only continent on Earth where the most overwhelming influence on climate is a nonannual, climatic change. The cycle driving Australia's unpredictable climate is called the **El Nino Southern Oscillation (ENSO)**. This cycle can cause drought one year and floods the next, or drought in one part of the country and floods in another.

Under normal climate conditions, a huge pool of warm water sits in the western pacific – around Indonesia, New Guinea and northern Australia. When an El Nino event occurs, the warm water begins to move eastwards towards South America, warming the water off Peru and Ecuador.

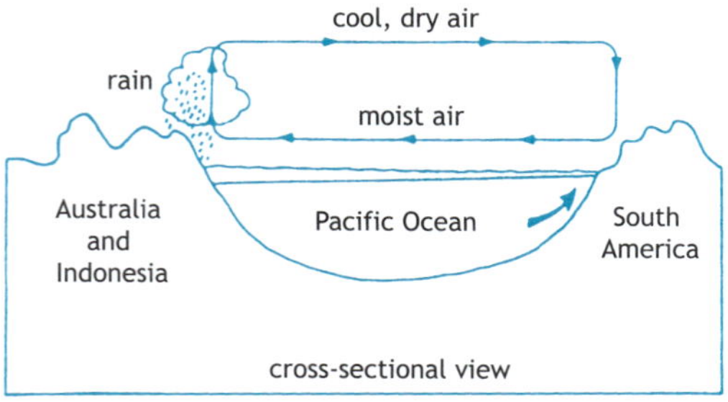

The normal climate situation

In Australia an El Nino event causes lower rainfall and droughts. In South America El Nino results in devastating floods and mudslides. This Southern Oscillation is now measured by recording sea surface temperatures using buoys across the Pacific. The Southern Oscillation Index (SOI) is a measure of the strength and phase of the Southern Oscillation. Negative values in the Southern Oscillation indicate an El Nino event, resulting in drought in Australia.

Glossary

Climate – The average weather conditions of a particular place.

Coriolis Effect – The deflection of winds and water by the spinning of the Earth – to the right in the northern hemisphere and to the left in the southern hemisphere.

Cyclone – A region of low atmospheric pressure accompanied by violent storm conditions.

El Nino – A term used to describe the movement of warm surface water from the western pacific to the coast of Peru and Ecuador.

Hectopascal – The international unit for measuring pressure.

High – A high pressure area where air molecules are close together.

Humidity – The amount of water vapour in the air at any one time.

Isobar – A line on a weather map joining points with the same air pressure.

Low – A low pressure area where air molecules are not close together.

Meteorology – The study of the atmosphere, its composition, and the changes that take place in it.

Precipitation – Any form of rain or snow. Relative humidity – The amount of moisture in the air relative to the maximum it can hold at that temperature.

Southern Oscillation – The seesaw of atmospheric pressures between the South Pacific Ocean and the Indo-Australian areas.

Southern Oscillation Index – Monitors the differences in sea level atmospheric pressures between Tahiti and Darwin.

Tornado – A whirlwind or mass of rotating air with high wind speeds at its centre.

Weather – The atmospheric conditions in a specific place at a specific time.

Wind – Air moving from a high pressure area to a low pressure area.

Science skills

Question 1

The table below displays the maximum and minimum temperature for several days at the beginning of February.

Day	Maximum temperature (°C)	Minimum temperature (°C)
1	30	22
2	29	21
5	31	21
6	30	22
7	29	21
8	30	22
9	24	21

a Draw a graph which displays this data.

b Which day has the highest maximum temperature?

c Which day has the largest difference between the maximum and minimum temperatures?

d Which day has the smallest difference?

Question 2

Use the compass diagram below to answer the questions.

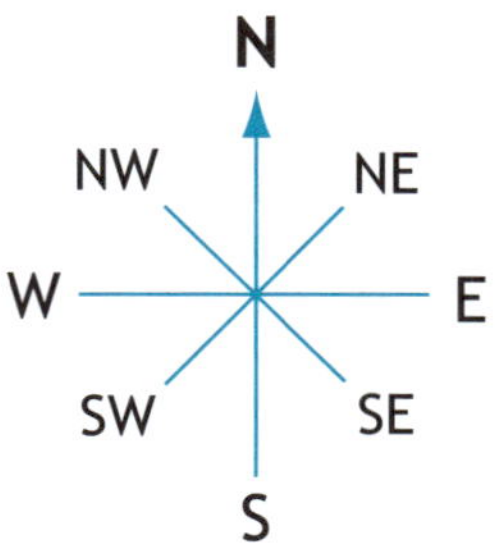

a Look at the wind sock below. From which direction is the wind blowing?

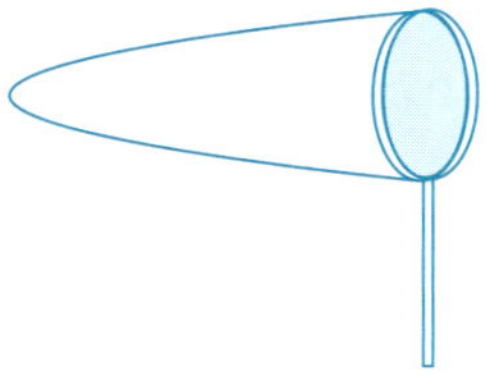

b What other instrument can be used to measure wind direction?

Question 3

The following graph displays the weekly rainfall data for the week ending 9 February 2001.

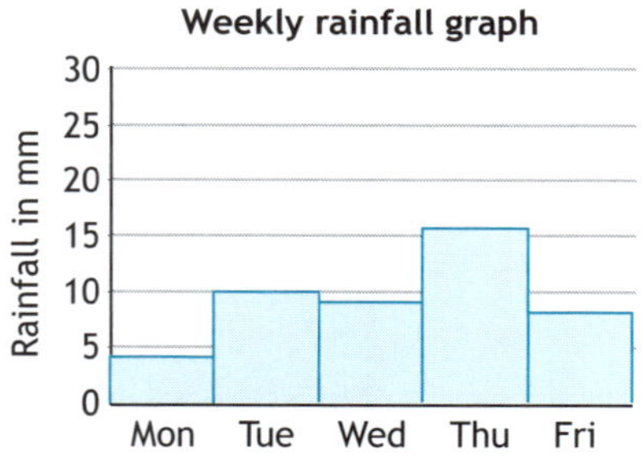

a How much rain fell on the Friday?

b Which day was the wettest?

c What is the average rainfall for the week?

Question 4

a Label the following diagrams with the terms below:

L, high pressure system, hectopascals, H, isobars, low pressure system

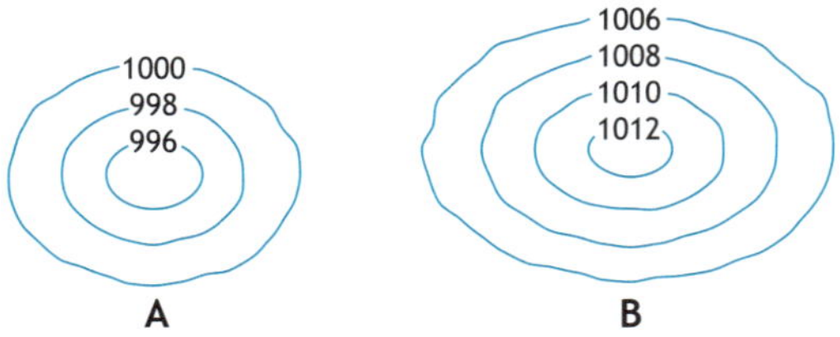

b Which diagram indicates fine weather and light winds?

c What does the term 'hectopascal' describe?

Question 5

Study the following weather chart displaying data for 7 July 2001.

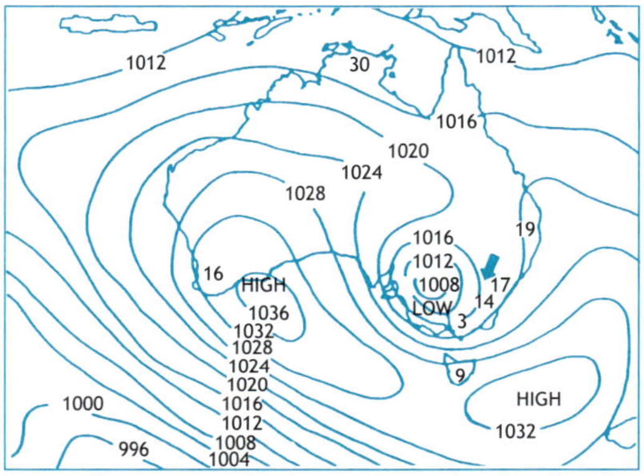

a What sort of weather are Victoria and southern New South Wales experiencing?

b Are the winds over Western Australia strong or gentle? How do you know this?

c In which overall direction are these pressure systems moving?

d What is the weather like in Townsville?

e Which direction are the winds in Perth coming from?

Question 6

a On the following diagram draw lines joining the places with equal air pressure.

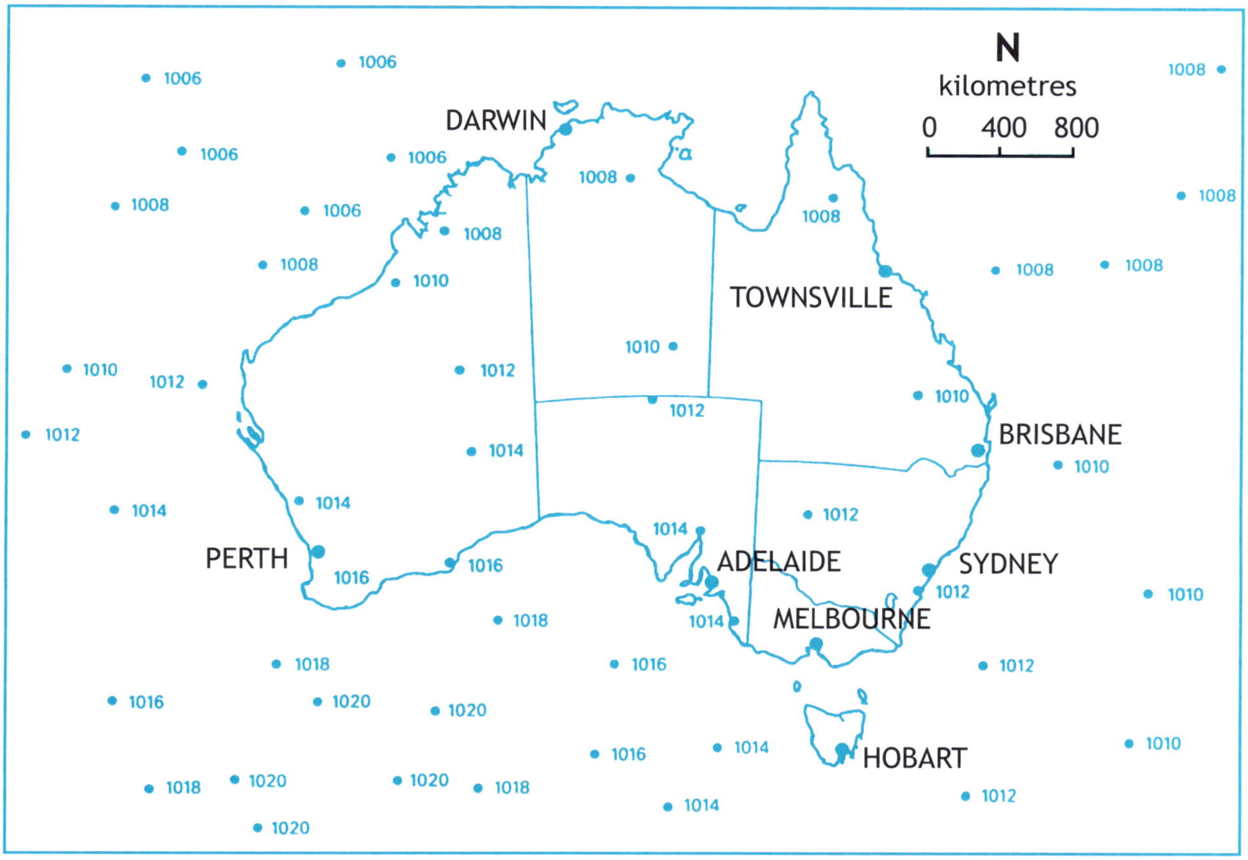

b Mark on the map any highs and lows.

c What is the weather like in Darwin?

Question 7

The following weather chart displays recorded data from a day in February.

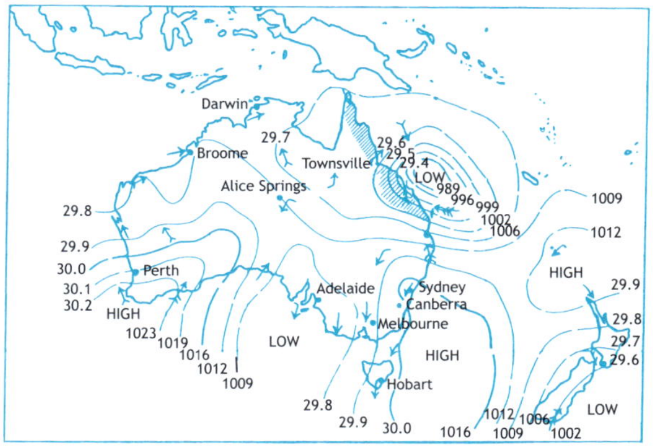

a What is the name of the severe weather pattern forming off the coast of Queensland?

b What does the shading indicate along the coast of Queensland?

c In which direction is this low pressure system rotating?

Question 8

The following diagram displays the Southern Oscillation Index from 1988 until 1998.

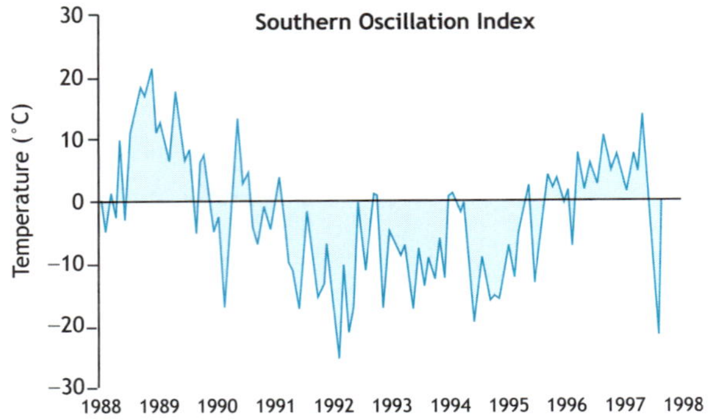

a Was 1989 a wet or dry year in Australia?

---

b What weather would you have predicted for 1998?

Question 9

Study the diagrams below. Which one shows an El Nino event?

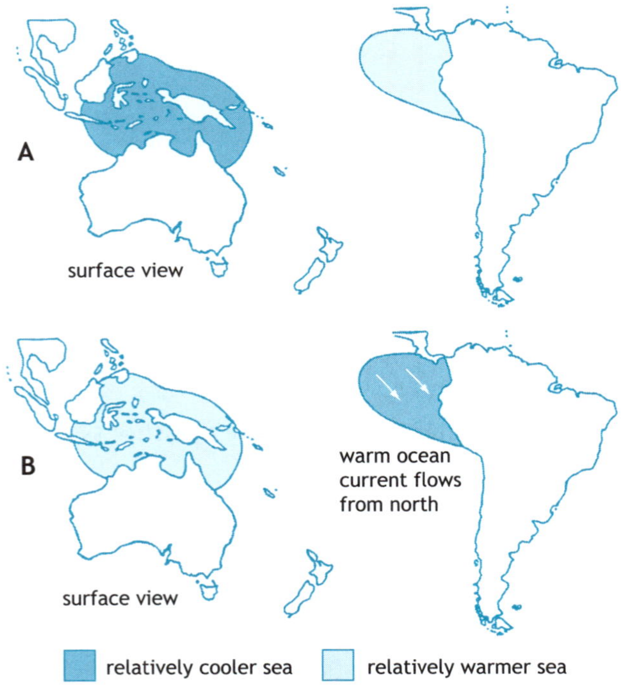

Question 10

Measuring wind speed can be done by using a protractor, string and a ping-pong ball. A student made this device to measure wind speed and used the Beaufort Wind Scale to describe the wind.

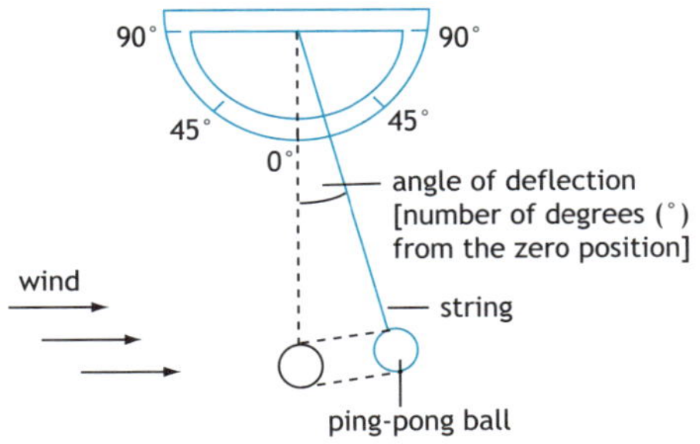

a Use the data in the table below to estimate wind speed for the following angles of deflection:

i 20° ______

ii 60° ______

iii 85° ______

String angle (degrees)	Wind speed (km/h)
90	90
85	09
70	19
60	24
55	26
30	42
20	52

b Use the Beaufort Wind Scale on page 98 to describe the wind and its effect for the above wind speeds.

Question 11

Humidity is the amount of water vapour in the air at a particular temperature.

To work out humidity:

- Measure the air temperature with a dry bulb (T1) and with a wet bulb (T2).

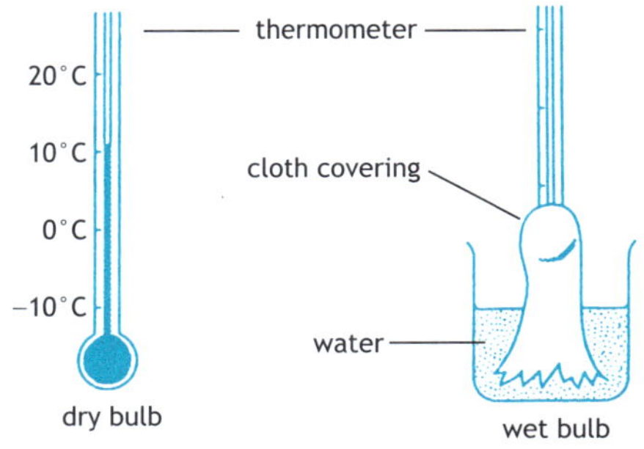

- Subtract T2 from T1 to calculate the difference.
- Use T1 and the temperature difference to work out the per cent humidity, using the table on page 98.

a If the humidity was 54% what was the reading on the wet bulb?

b If the temperature difference was 3°C and the dry bulb was 16°C, what was the humidity?

c What is the humidity when the dry and wet bulbs show the readings below?

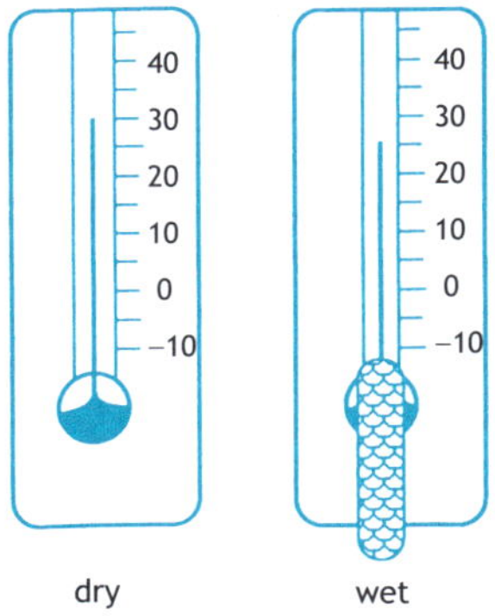

Question 12

The following diagram shows typical variations in atmospheric pressure and wind across a severe weather pattern. These patterns often form over tropical waters and can cause widespread devastation.

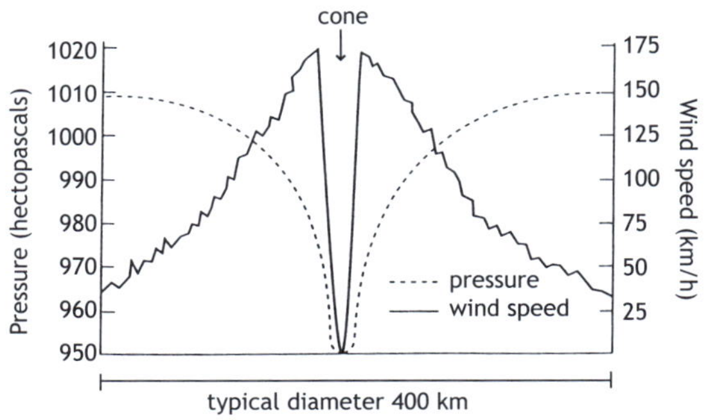

a What is the name of this weather system?

b What is the wind speed at the edge of this system?

c What is the wind speed around the central core and inside the core?

d What is the atmospheric pressure in the centre of the system?

Question 13

The data below represents the rainfall data for several Australian towns.

City	Jan	April	July	Oct
Adelaide	20	44	66	44
Brisbane	164	87	57	76
Alice Springs	25	14	15	22
Darwin	428	99	1	72
Hobart	48	53	53	63
Perth	8	45	174	55
Sydney	102	124	101	78

a Which town has the highest rainfall in summer?

b Which town has the driest winter?

c Which town has consistently low rainfall?

____

d Which town receives mostly winter rainfall?

____

Question 14

The following graphs display the frequency of thunderstorms in four Australian cities.

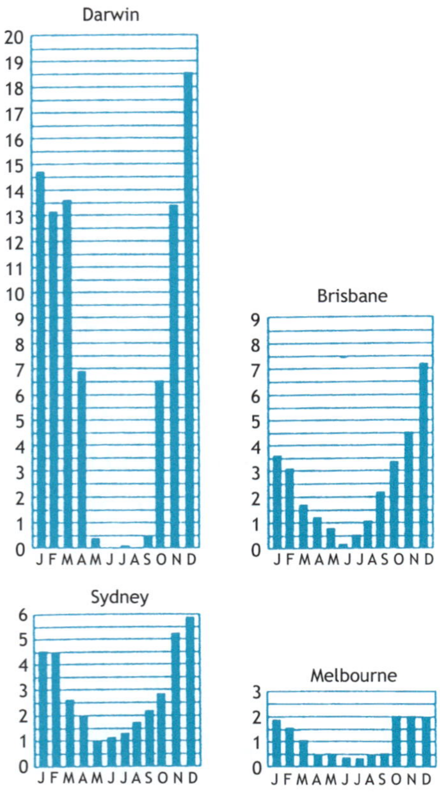

a What is the relationship between thunderstorm activity and distance from the equator?

____

____

b In Darwin, which month has the greatest thunderstorm activity?

____

c What season generally has the lowest risk of thunderstorm activity?

____

Revision questions

Question 1

Complete the following information by filling in the correct words.

Clouds are visible when ________ ________ in the form of water droplets is present in the air. Clouds are classified depending on how ________ they are in the atmosphere. The three main forms of clouds are ________, ________ and ________.

Question 2

Match the following terms with the correct explanation.

Climate	The study of the atmosphere.
Weather	Average weather conditions of a particular place.
Meteorology	Atmospheric conditions in a particular place at a particular time.

Question 3

Weather is recorded daily using various instruments. What are the following instruments used to measure?

a barometer

____

b rain gauge

____

c anemometer

____

d maximum / minimum thermometer

____

e wind vane / wind sock

____

Question 4

a What unit is used to measure temperature?

b On a weather map how are areas of rain marked?

c What is the Beaufort Wind Scale?

d What additional information can be gained by knowing the direction of the wind?

e What does a falling barometric reading mean?

Question 5

Match the following terms with the correct description.

Thunderstorm	A narrow, violently rotating column of air.
Cyclone	A massive, spiralling low pressure system causing widespread devastation.
Hurricane	A storm accompanied by heavy rain, gusty winds and lightning.
Tornado	A region of low atmospheric pressure accompanied by violent storm conditions.

Question 6

Explain the meaning of the following terms.

a ENSO

b SOI

c El Nino

d negative SOI/positive SOI

Question 7

Air's capacity to hold water depends on the temperature. The graph below displays this relationship.

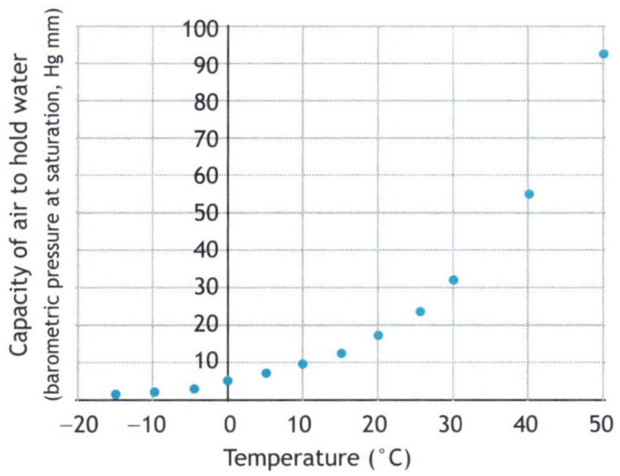

a Complete the following information by adding the correct words.

The __________ the water vapour, the ______ it can hold. When the maximum amount of water vapour is present at a certain temperature and the air is ________, the water vapour will begin to __________ around __________ to form water droplets. Billions of droplets form ________.

b Explain the term 'precipitation'.

c How is hail or snow formed?

Question 8

The following diagram represents a tropical low. Is the depression in the northern or southern hemisphere? Explain your answer.

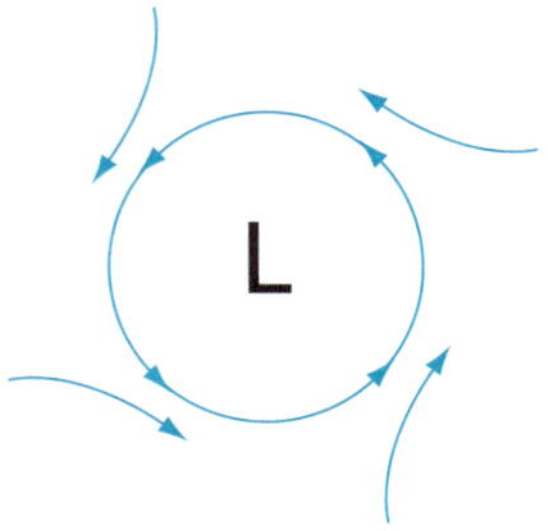

Question 9

a What is the name given to air when it starts moving?

b Air tends to move either vertically or horizontally. In each case, what causes it to move?

c What is the name given to the effect of the Earth's rotation on air movement?

Question 10

a What force holds the atmosphere close to the surface of the Earth?

b What term do we use for the weight of air?

c Which weighs the most – cold air or warm air?

d A low pressure system would generally be associated with cloudy, rainy weather or fine weather?

Question 11

a Study the following diagram and then label all the highs and lows.

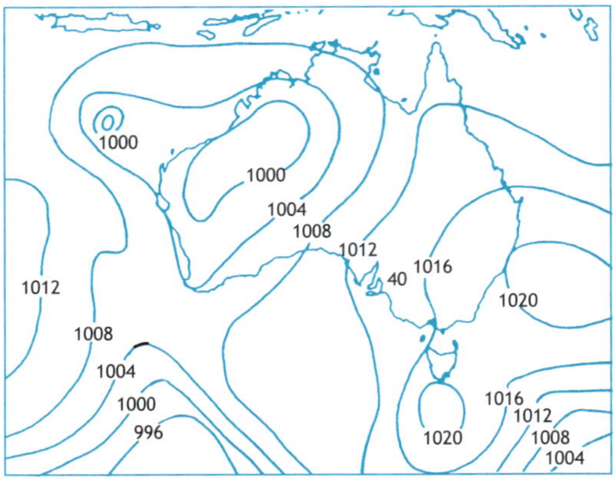

b What is the lowest pressure recorded here?

c What is the air pressure in Melbourne?

d What instrument is used to measure air pressure?

e What name do we give the lines that connect up points of the same pressure?

f What type of weather is indicated by these lines when they are very close together?

Question 12

The following chart shows the areas of Australia affected by a negative El Nino event in 1992.

Choose the correct answer.

The shaded area represents:

a extremely high temperatures.

b extremely low rainfall.

c higher than average rainfall.

d colder than average temperatures.

Beaufort Wind Scale

Beaufort number	Description	Wind speed (km/h)	Effect
0	calm	less than 2	smoke rises vertically
1	light air	2 – 5	smoke drift shows wind direction; wind vanes don't move
2	light breeze	6 – 12	wind felt on face; wind vanes move
3	gentle breeze	13 – 20	leaves and small twigs in motion; hair disturbed; clothing flaps
4	moderate breeze	21 – 30	dust and loose paper moved; small branches move
5	fresh breeze	31 – 40	small trees with leaves begin to sway; wind force felt on body
6	strong breeze	41 – 51	large branches move; umbrellas difficult to use; difficult to walk steadily
7	moderate gale	52 – 63	whole trees in motion; inconvenience felt when walking
8	gale	64 – 77	twigs broken off trees; difficult to walk
9	strong gale	78 – 86	people blown over; slight structural damage, including tiles being blown off houses
10	whole gale	87 – 101	trees uprooted; considerable structural damage
11	storm	101 – 120	widespread damage
12	hurricane	greater than 120	widespread devastation

The humidity of air

Dry bulb (°C)	Difference between dry bulb and wet bulb readings (°C)																				
	0	1	2	3	4	5	6	7	8	9	10	11	12	13	14	15	16	17	18	19	20
0	100	82	65	48	31																
2	100	84	68	52	37	22															
4	100	85	70	56	42	29															
6	100	86	73	60	47	35	23														
8	100	87	75	63	51	40	29	18													
10	100	88	76	65	54	44	34	24	14												
12	100	89	78	68	57	48	38	29	20												
14	100	90	79	70	60	51	42	33	25	17											
16	100	90	81	71	63	54	46	37	30	22	15										
18	100	91	82	73	65	56	49	41	34	27	20										
20	100	91	83	74	66	59	51	44	37	30	24	18									
22	100	92	84	76	68	61	54	47	40	34	28	22	16								
24	100	92	84	77	70	63	56	49	43	37	31	26	20	15							
26	100	92	85	78	71	64	58	52	46	40	34	29	24	19	14						
28	100	93	85	78	72	65	59	53	48	42	37	32	27	22	18						
30	100	93	86	79	73	67	61	55	50	44	39	34	30	25	21	17					
32	100	93	86	80	74	68	62	57	51	46	41	37	32	28	24	20	16				
34	100	94	87	81	75	69	64	58	53	48	43	39	34	30	26	22	19	15			
36	100	94	87	81	76	70	65	60	55	50	45	41	37	32	29	25	21	18	14		
38	100	94	88	82	76	71	66	61	56	51	47	43	39	35	31	27	24	20	17	14	
40	100	94	88	82	77	72	67	62	57	53	48	44	40	37	33	29	26	23	20	17	14

CHAPTER 6

FORCE AND MOTION

Why do cars have airbags?

Sir Isaac Newton (1643–1727) was an English physicist and mathematician who was very interested in force and motion. He studied closely what happened as objects moved. His three laws of motion are still used by scientists today to understand the motion of anything from tiny atoms to planets.

What is a force?

Whenever we are lifting, bending, pushing, pulling, twisting, tearing, stretching or squeezing we are exerting a **force**.

Forces can:

- change the speed of an object.
- change the direction of movement of an object.
- change the size or shape of an object.

Measuring forces

Forces are measured in units called **newtons (N)** named after Sir Isaac Newton. **Weight** is a force. It is due to the pull of gravity on the Earth. A spring balance uses a spring to measure the weight of an object or the strength of any pulling force. The **mass** of an object is the amount of matter in it. This is measured in kilograms. A mass of 1 kg on Earth weighs almost 10 newtons. A 1 kg weight in space would be weightless, but its mass would be the same as on Earth.

Balanced forces

A force is needed to start something moving or to change its direction, however, forces are also acting on things that are and not moving. If something is staying still the forces acting on it must be balanced.

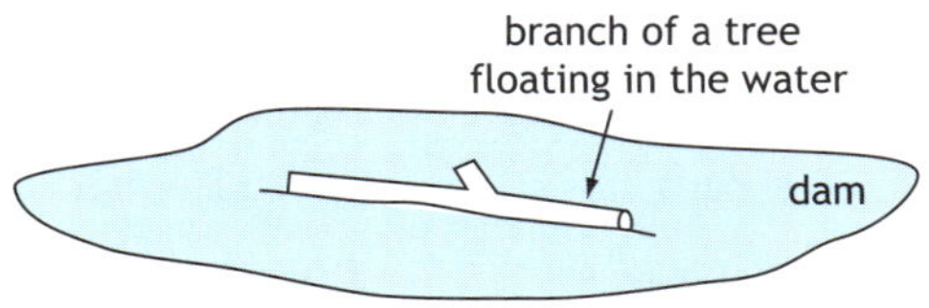

Balanced forces

The forces acting on a branch floating in a dam are balanced. The weight of the branch is balanced by the water pushing up. A force doesn't always change the way something moves because the force may be balanced by another force. To change the way something moves, you need an extra force.

Newton's First Law of Motion: Inertia

When an object is stationary, it needs a force to make it move. The bigger the mass the bigger the force needed to make it move. The reluctance of an object to start moving is called **inertia**. Since inertia depends on the mass of an object, inertia will be the same on the Earth as it is in space — if you push a large mass in space you feel the same amount of inertia as on Earth.

Moving objects also need a force to stop them moving, as their inertia tends to keep them moving in the same direction.

For example, why do you feel pressed back into the seat when a car takes off quickly? Because your inertia tries to keep you in the same place, it takes some effort to get your body moving at the same speed as the car. Why are you thrown forward when a car stops suddenly? Because your inertia tries to keep you moving forward and it takes the force from your seatbelt to slow you down.

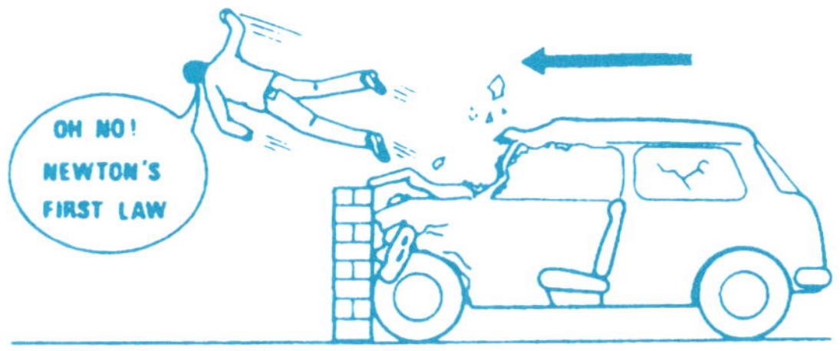

Newton's First Law

Sir Isaac Newton stated all this in his First Law of Motion — Every mass stays at rest or moves at a constant speed in a straight line unless a resultant force acts on it. In other words:

If the forces on a mass are balanced then:

- if it is at rest, it stays at rest.
- if it is moving it keeps on moving at a constant speed in a straight line.

Turning forces – circular motion

Some objects won't move along because they are fixed to the ground, but they may move around a central point.

If you whirl a cork on a piece of string around your head, you can feel the force in the string.

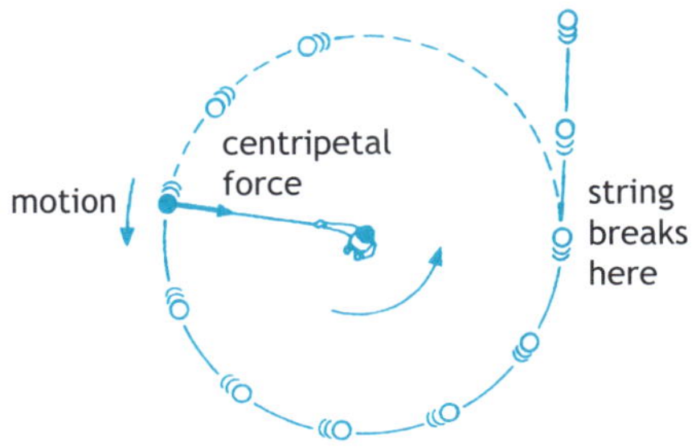

Centripetal force

Newton's First Law says that the cork will continue to move in a straight line unless a force is applied. The force is your hand pulling on the string. This force is acting towards the centre of the circle and keeps the cork from going straight ahead. This force is called centripetal force. It is used to tighten a bolt, spin-dry clothes or close a door.

Newton's Second Law of Motion

If you want to move an object to another place you will have to add a force. Two things about the force are important:

- the **size** of the force; and
- the **direction** of the force.

Because both size and direction are important force is called a **vector quantity**. Vector quantities have both direction and magnitude. **Velocity** and **acceleration** are both vector quantities.

The Italian scientist, Galileo (1564 – 1642), was interested in force and motion and carried out many experiments on velocity and acceleration. Galileo's ideas laid the groundwork for Sir Isaac Newton, who drew the ideas together into his three laws.

Resultant forces

- Vector quantities like forces can add up. For example forces of 2 N and 2 N acting in the same direction add up to a resultant force of 4 N.
- Forces can be taken away — forces of 2 N and 2 N acting in opposite directions cancel out and there is no resultant force. So the resultant force is zero. These are balanced forces — Newton's First Law.
- So if a 5 N force was opposed by a 2 N force the resulting force would be 3 N.

Velocity

Velocity is almost the same thing as speed but it is a vector quantity so it has size (speed) and direction.

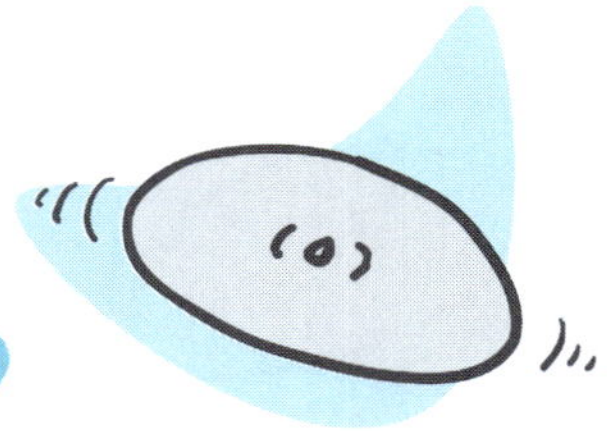

If a bicycle is moving at a steady speed of 10 metres every second, its speed is not changing it has a **constant speed**.

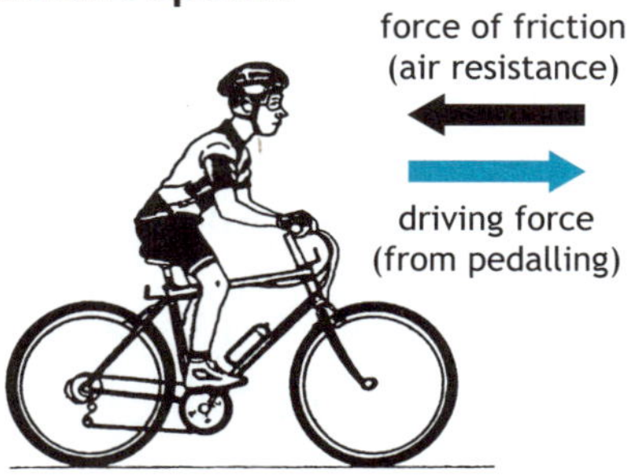

Balanced forces – constant speed

If it is moving at a constant speed in a straight line it has **constant velocity**. A spacecraft moving through space in the absence of forces has a constant velocity.

Acceleration

Acceleration is just a change in velocity. If the person on the bicycle pedals faster the bicycle goes faster — it accelerates. If the bicycle accelerates the velocity increases.

- An object accelerates (or decelerates) only if a force is applied to it.
- Forces control acceleration but not velocity. A push or pull makes things accelerate but they do not affect velocity directly.
- There is no real difference between something moving at a steady rate and some that is not moving at all.
- Acceleration depends on the size of the force and how much mass the object has. A large force accelerates a light object very fast, a small force accelerates a massive object slowly.

Newton's Second Law of Motion states that when an unbalanced force acts on an object it accelerates in the same direction as the force. The acceleration is directly proportional to the force and inversely proportional to the mass.

In other words:

Force (N) = mass (kg) × acceleration (m/s/s)

$F = ma$

Where:

- F is the resultant (or unbalanced force) force on the mass (m)
- The force (F) must be measured in newtons.

One newton (1 N) is the force which gives to a mass of 1 kg an acceleration of 1 m/s/s.

Example

What is the force of gravity on a 1.0 kg mass if the acceleration is 9.8 m/s/s ?

$F = ma$

$F = 1 \times 9.8$

$= 9.8$ N

A force is an action exerted upon a body, in order to change its state of rest or its state of uniform motion in a straight line. Action tends to produce a change in motion.

Newton's Third Law of Motion

Friction is a force which opposes the movement of an object. Friction is often a nuisance, because it converts kinetic energy into heat and wastes it.

The friction between moving objects can be reduced by:

- using smooth surfaces.
- lubricating surfaces with oil.
- separating the surfaces with air (eg hovercraft).
- use wheels or rollers.
- use streamlined shapes.

Friction can also be very useful, for example in the brakes on cars and bicycles, and the air friction of a parachute.

Newton noticed that if two teams were having a tug of war and the rope was not moving, then the forces must be balanced. Team A is pulling with the same force as Team B.

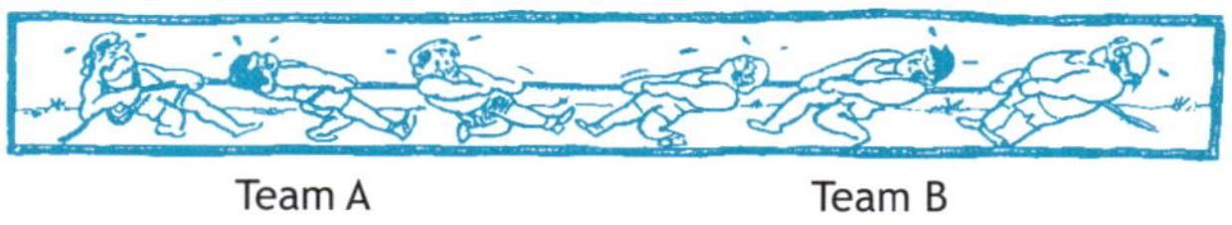

Newton observed that forces are in always in pairs and that the two forces are always equal in size but opposite in direction. He called the two forces **action** and **reaction**.

In his **Third Law**, Newton stated that: For every action there is an equal and opposite reaction.

In other words:

> The action force and the reaction force are equal and opposite.

For example, if you rest your elbow on a table:

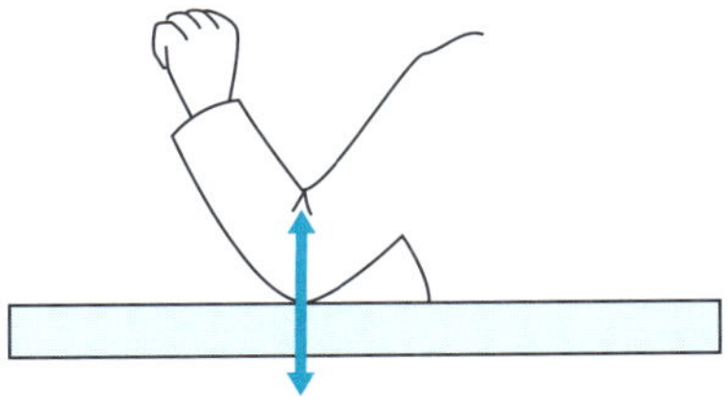

Equal forces

the force of your elbow on the table (downwards)	=	the force of the table on your elbow (upwards).

Spacecraft provide another example:

As the fuel inside a spacecraft burns, hot gases expand and then rush from the rocket at tremendous speed through the only opening – the exhaust nozzle. The force of the gases rushing in one direction sends the rocket in the opposite direction.

Newton's three laws have given scientists a good understanding of the way force and motion are related, as well as a way of analysing them mathematically. Until the 1920s they seemed to explain every movement in the universe perfectly. Physicists are now finding even deeper understandings of force and motion.

Glossary

Acceleration
The rate of change of velocity (unit = m/s/s).

Constant velocity
An object moving in a certain direction at a constant speed (unit = m/s).

Force
Something which changes the shape or motion of an object – stretching, squashing, pulling, pushing (unit = newton).

Friction
The force that resists movement when one surface slides against another.

Inertia
An objects resistance to change in motion, depending on its mass.

Mass
The amount of matter in an object, measured in grams (g) or kilograms (kg).

Newton
The unit that we use to measure forces.

Vector quantity
Quantity with direction, such as velocity.

Velocity
Speed in a particular direction.

Weight
The force of gravity that pulls an object towards the Earth.

Question 1

How much would 1 kg weigh on the Moon if the pull of gravity there is about one-sixth of the pull of gravity on Earth?

____

Question 2

A student decided to find out some more about inertia. He tied two tins from two long pieces of string. He filled one tin with wet sand (A) and the other was empty (B).

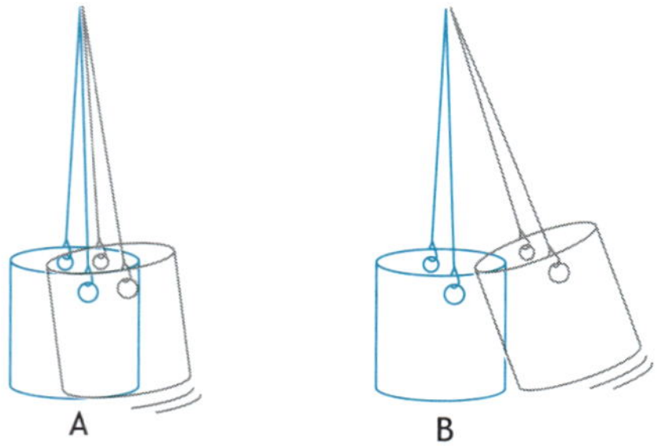

a Which tin is harder to push? ________

b Which tin has the greatest inertia? ________

c Which tin has the larger mass? ________

d If both tins are swinging which tin is harder to stop? ________

Question 3

Study the drawing below.

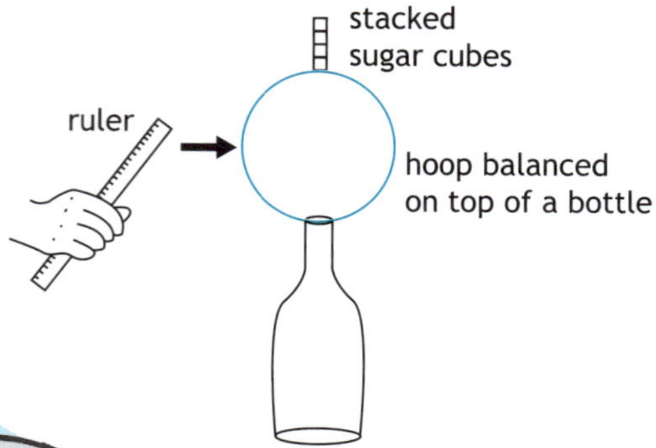

a What will happen to the sugar cubes if the hoop is quickly knocked aside?

____

____

b What stops the sugar cubes from scattering as the hoop moves?

____

____

Question 4

Complete the following table by balancing the forces.

Situation	Force one	Force two
person holding a briefcase		
boat floating		
hot air balloon tethered to the ground		

Question 5

Consider the block of wood resting on the table below.

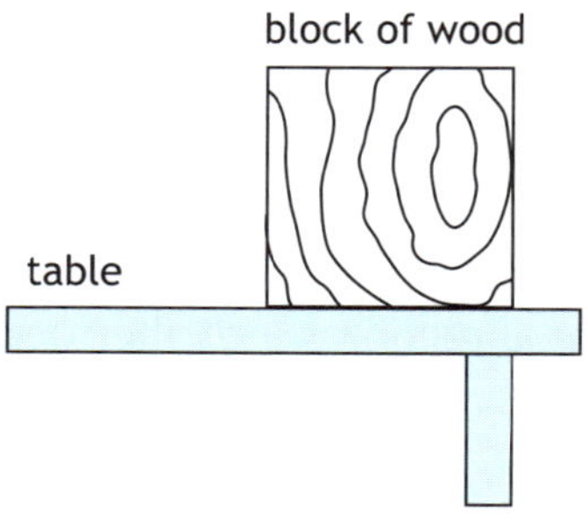

a Where is the pair of equal forces?

____

____

b Which of Newton's Laws is being applied here?

____

Question 6

Two students tried a simple test to find out about Newton's Third Law of Motion. They faced each other, each standing on a skateboard. They put the palms of their hands together and pushed.

a What do you predict happened?

____

____

b How does this result support Newton's Third Law?

____

Question 7

Study the drawing below of a parachutist who is falling at a steady speed.

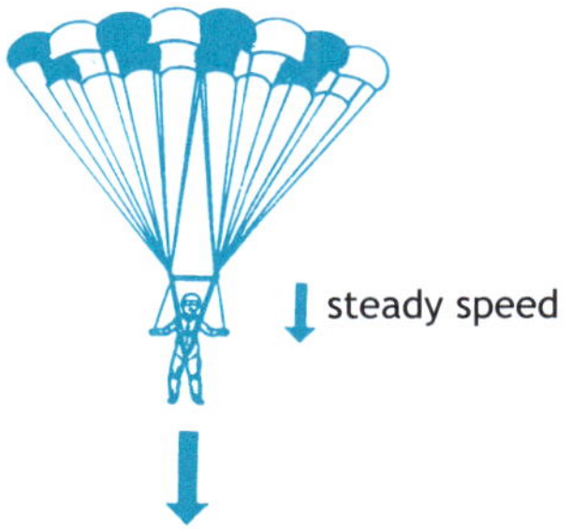

a Using arrows, indicate the direction of the forces acting on this person.

b Beside each arrow write the name of the type of force acting.

c Why does the parachutist fall at a steady speed?

____

Question 8

The diagram below is of a helium balloon which is not moving up or down.

Explain how forces are keeping the balloon steady.

____

____

____

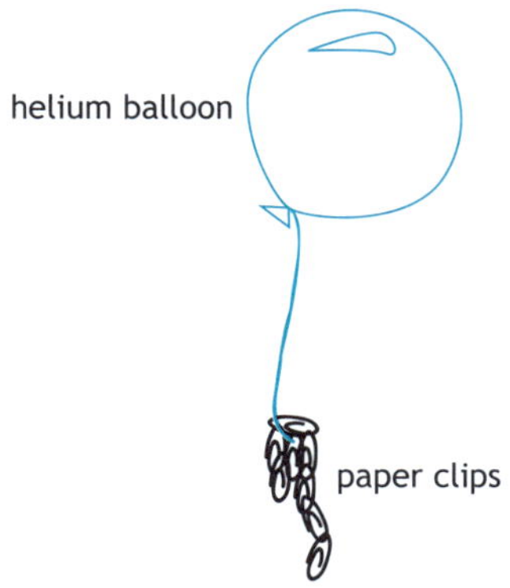

Question 9

Study the car moving along a horizontal road at a constant velocity. Vector X is the driving force on the car (supplied by the motor). Vector Y is the sum of all the retarding forces acting on the car (such as friction, air resistance).

a Which of the following gives the correct relationship between the sizes of these two forces?

A X is smaller than Y.

B X is bigger than Y.

C X is equal to Y.

D None of the above.

Question 10

A group of students set up the following experiment. They opened a clothes peg and tied it so it stayed open. They put a small marble on one side and a large on the other side then cut the string.

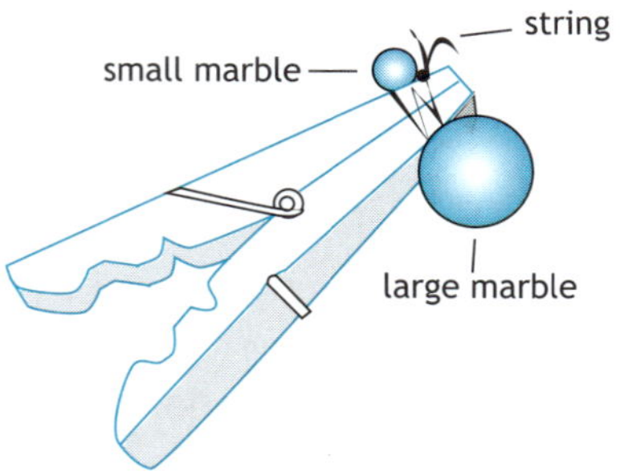

a Which marble travelled the furthest?

____

b Which of Newton's laws does this result support?

____

Question 11

The experiment below was set up by a teacher.

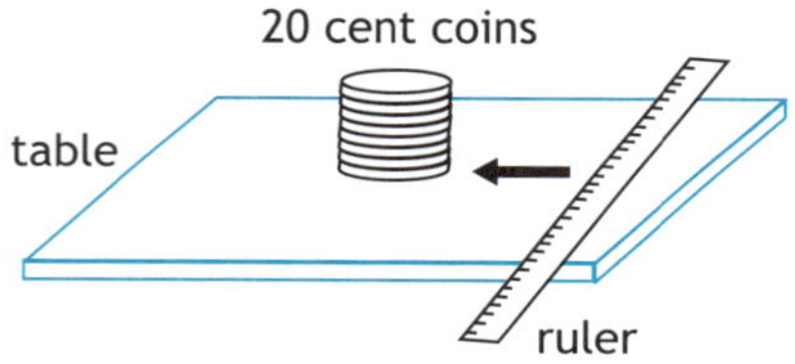

He slid the ruler rapidly along the table and knocked the bottom coin out. The rest of the coins stayed in position. Then he repeated this action for all the coins.

Explain what the teacher was doing here in terms of Newton's Laws.

____

____

____

____

Question 12

What is the resultant force acting on the following objects:

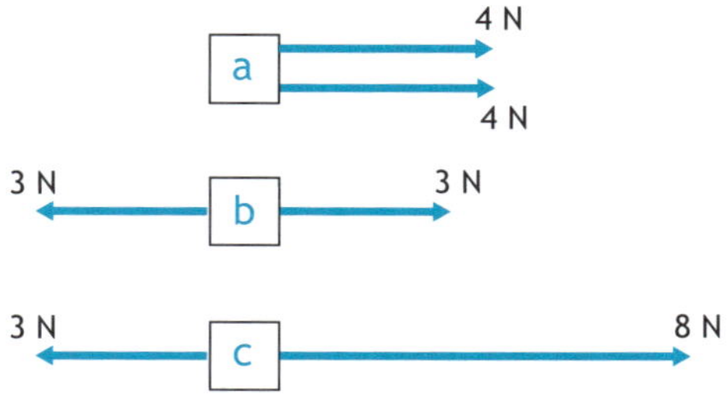

Question 13

a Should this person pull the tablecloth quickly or slowly?

____

b Explain the physics of this trick.

____

____

____

Question 14

A parachute is designed to have a lot of air resistance. As the skydiver falls he transfers his gravitational potential energy into kinetic energy. Look at the speed-time graph for the skydiver.

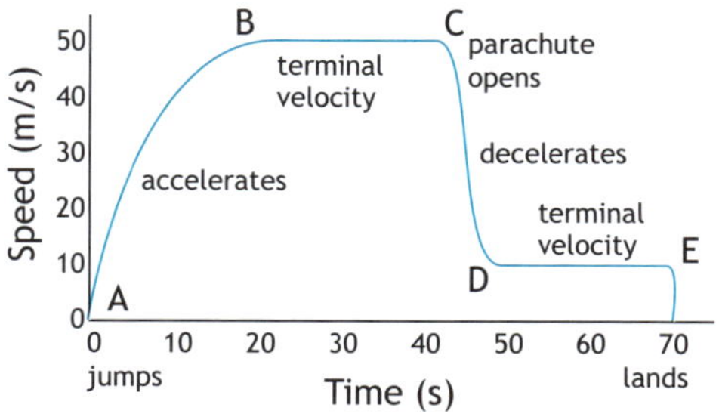

a What is his speed at point A? __________

b What is his speed at point B? __________

c How long does he take to accelerate to terminal velocity? ____________

d What happens to his speed when his parachute opens?

____

e How long does it take to decelerate to terminal velocity?

____

Question 15

A skydiver weighing 700 kg jumps out of a helicopter and uses air resistance to land safely.

Match the drawings with the correct explanation and then put them in the correct order.

A

(1) The parachute opens and the two forces (friction and weight) become equal again

B

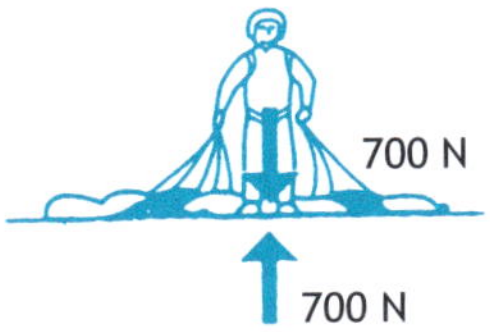

(2) There is only one force on the skydiver (weight).

C

(3) The resultant force (air resistance) is now upwards making him slow down.

D

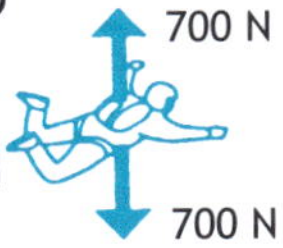

(4) As he free falls the forces (friction and weight) become equal, so he stops accelerating and falls at a constant speed.

E

(5) When he hits the ground the Earth pushes up and he decelerates fast.

Question 16

If the trolley is pulled forward the ball appears to roll back and hit the back of the trolley. Explain what is really happening.

____

____

____

____

Question 17

Explain the physics of this cartoon.

____

____

____

____

Question 18

Explain what forces are acting when you begin to walk or run?

____

____

____

Question 19

Which of Newton's Laws does this diagram explain?

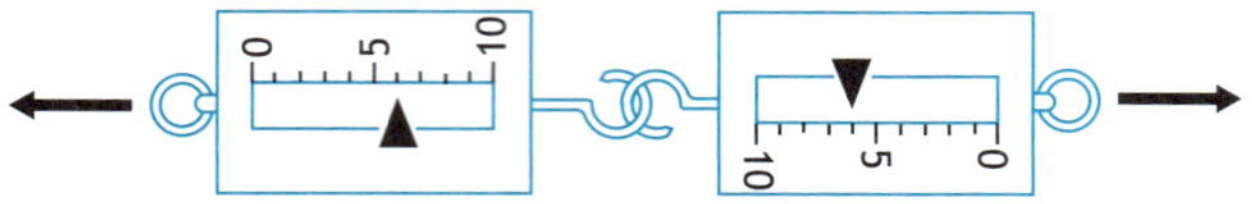

Question 20

What is the centripetal acceleration on a 2 kg mass if the acceleration is 10m/s/s?

Question 1

Study the diagram of a balloon rocket below and put the correct words into the explanation.

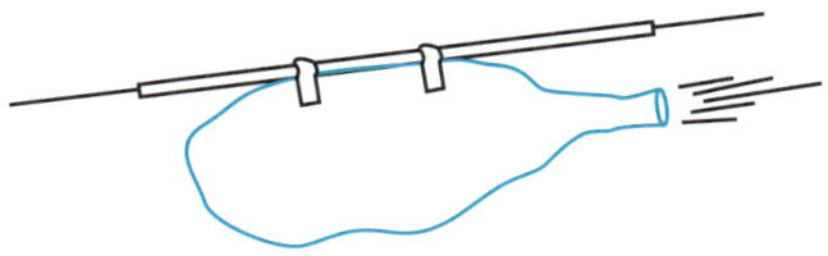

When the air inside the balloon escapes, it pushes the balloon ________. The principle on which the balloon rocket works is Newton's ________ Law of Motion: for every ________, there is an equal and opposite ________.

Question 2

If a driver brakes and the car decelerates steadily:

a What provides the force that decelerates the car?

b If the deceleration is constant, what can you say about the force?

c If the passenger is to remain in his seat he must have the same deceleration as the car. What provides the force under normal circumstances?

d What happens if the passenger is not wearing a seatbelt and the driver brakes suddenly?

Question 3

a If you pushed both supermarket trolleys with the same force, which trolley would be the easiest to get going? Why?

b Which trolley is the easiest to stop? Why?

c If you took both trolleys into outer space, would they weigh the same?

d While out in space, if we pushed both trolleys, would they need the same sized force to accelerate?

Question 4

a What is a newton?

b Does a newton of lead weigh more or less than a newton of feathers?

Question 5

Car bodies are designed so that airflow is as smooth as possible. They are streamlined to reduce the air resistance caused by friction.

The size of the air resistance depends on:

a the shape of the object.

b the size of the object.

c the speed of the object.

d all of the above.

Question 6

If you are standing on a bus, what happens to you when the bus:

a starts moving?

b stops moving?

c turns a corner to the left?

d Which of Newton's Laws explains these?

Question 7

Fill in the missing words.

Forces can:

a change the s_ _ _ _ _ of an object;
change the d_ _ _ _ _ _ _ _ of
m_ _ _ _ _ _ _ _ of an object;
change the s_ _ _ or s_ _ _ _ of an object.

b ________ is the force due to the pull of gravity. ______ is the amount of matter in an object. Force is measured in ______ (N) and mass in __________ (kg).

c ________ is the reluctance of a mass to change its motion.

Question 8

An astronaut has a mass of 60 kg.

a What is her weight here on Earth? _______

b What is her weight on the moon? _______

c What is her mass on the moon? ________

Question 9

What does Newton's First Law of Motion state?

Question 10

Choose from the list below to explain what the force is doing.

- Bending an object.
- Balancing another force, and preventing movement.
- Stopping an object that is moving.
- Starting an object moving.
- Changing the direction of movement.
- Stretching an object.

Question 11

When a tennis player hits a tennis ball, what forces are acting on the ball as it moves across the net?

a The force of gravity.

b The force of friction.

c The force of the hit.

d No forces.

Question 12

The bigger the acceleration the bigger the force needed to produce it. The larger the mass of an object the larger the force needed to accelerate it. Use Newton's First Law and this information to answer the following questions.

Students suspended two pieces of thread about 3 metres long from a horizontal support. Both threads were wrapped around identical paperback books so that they hung down the same distance.

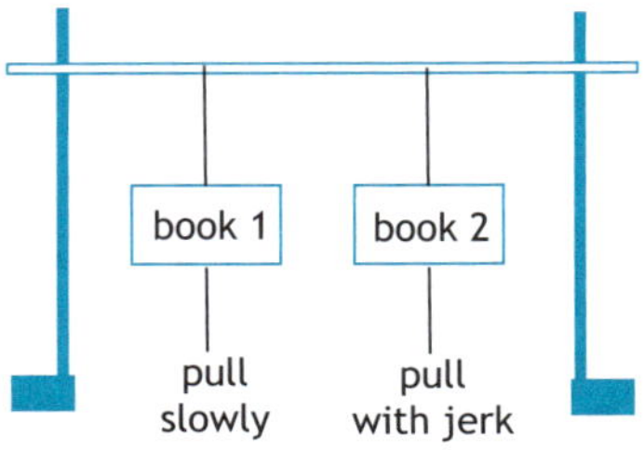

a What will happen if a student pulls slowly on the bottom thread of the first book? Why?

____

____

b What will happen is the student pulls sharply on the bottom thread of the second book? Why?

____

____

Question 13

Explain what is happening when the card is knocked quickly away from under the coin.

____

____

____

Question 14

State Newton's Third Law of Motion.

____

____

Question 15

Complete the boxes in the table below with the correct explanation.

1. Speed in a particular direction.
2. The unit we use to measure forces.
3. An object's resistance to a change in motion.
4. The rate of change of velocity.
5. The force of gravity that pulls an object towards Earth.

Term	Explanation
weight	
acceleration	
inertia	
velocity	
newton	

Question 16

Students set up a catapult like the one drawn below.

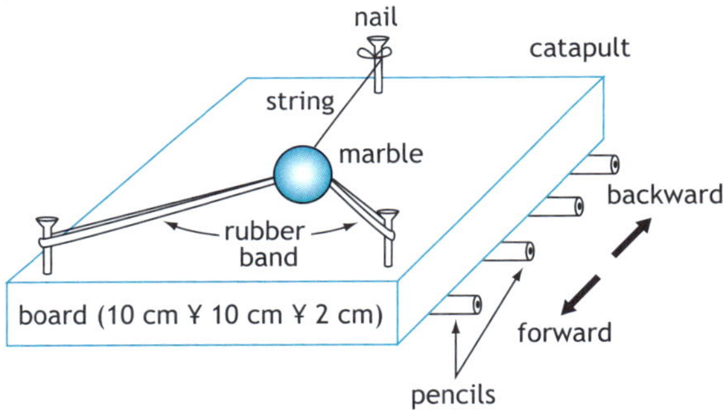

a If the string is cut which way does the catapult travel?

b Which way does the marble travel?

c If the pencils were removed, would the catapult move as much?

d Does the marble travel more or less when the pencils are removed?

Question 17

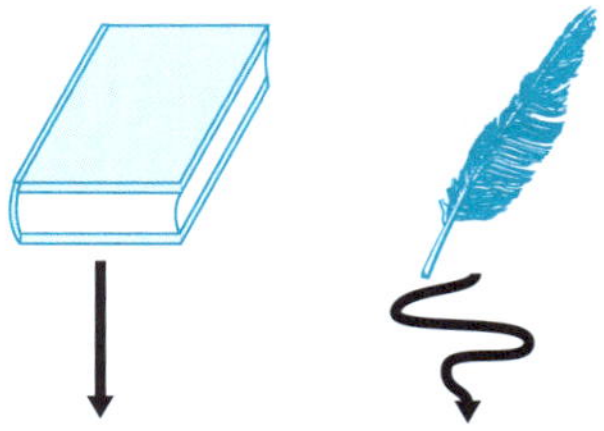

a A feather and a book are dropped at the same time from the same height. Which one hits the ground first?

b If the feather is placed on top of the book and both are dropped together what is the result? (3 lines)

Question 18

a A child is standing on a stationary skateboard. He steps off the skateboard, causing it to roll. Which moves the greater distance, the child or the skateboard?

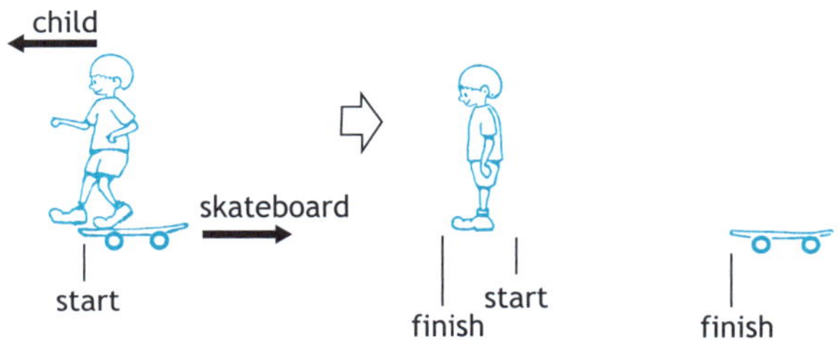

b Two children, one big and one small, each stand on a skateboard, facing each other with hands together. If they push each other apart which child moves the greatest distance?

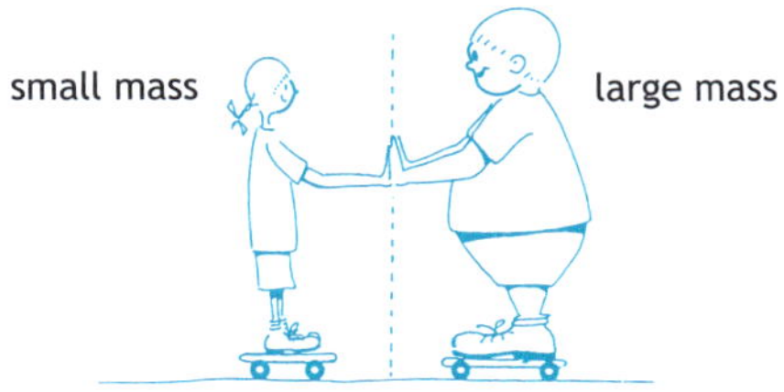

CHAPTER 7

ENERGY RESOURCES

What do squashed forests become?

Our daily lives depend on vast amounts of electricity. The modern world exploits energy to produce this electricity. Over 80% of the world's energy comes from burning **fossil fuels** — oil, natural gas and coal — which were formed many years ago from the fossilised remains of plants and animals. We also use huge amounts of fossil fuel to support our transport industry. We are burning up these fuels at an increasing rate, but they cannot be replaced and will eventually run out. These fuels are called **non-renewable** energy resources.

To provide fuel for the future we need to become more energy efficient and develop the use of more sustainable sources of energy such as solar, wave, thermal and water. These forms of energy are called **renewable** energy sources. Most of our energy, whether it be renewable or non-renewable, comes directly from the sun. Nuclear fusion reactions in the heart of the sun turn hydrogen into helium, releasing huge quantities of energy mainly in the form of light and heat.

Non-renewable energy resources

The burning of fossil fuels releases energy which has been stored as chemical energy. This energy can be released in the forms of heat, sound and light. Fuels themselves won't burn unless they have a supply of heat and oxygen.

Oil

It took 200 million years for the Earth to create the world's oil reserves, and we only have about 30 years' supply left. In Australia we only have about 10 years' worth of our own oil left. When it runs out we will have to buy oil from other countries and it may be too expensive.

Transporting oil is hazardous to the environment. Oil tankers sometimes spill the oil into the sea, resulting in the death of thousands of sea birds and millions of fish, and coastlines being fouled.

We rely on oil products for nearly all our transport. Petroleum or oil is the starting material for petrol, diesel, kerosene, LPG, bitumen and many plastics. About 90% of petroleum production is used for transport. Road transport uses four times as much energy as rail, air and sea transport combined. Unfortunately, Australians rely heavily on road transport.

Coal

Australia has lots of coal, which is the most commonly used fuel for **electricity generation**. The coal is burnt and the heat produced is used to generate electricity. The conversion is not very energy-efficient — only 35% of the heat energy is transformed into electrical energy; the other 65% being lost to the environment.

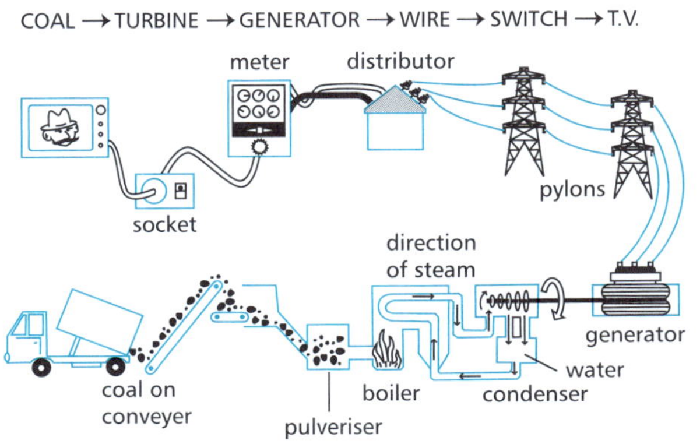

A coal-fired generator

The burning of coal and oil produces **exhaust gases**, such as carbon dioxide and sulphur dioxide, which affect the atmosphere. Problems such as the greenhouse effect, acid rain and smog are arising as a result of the burning of fossil fuels.

Several countries have now committed themselves to reducing their production of greenhouse gases.

Natural gas

In Australia we may only have enough natural gas reserves for another 60 years. We mainly use gas for cooking and to heat our homes, but by comparison with other nations, we make relatively little use of natural gas.

Nuclear power

Nuclear power, fuelled by uranium, once seemed to offer a cheap, clean, long-term alternative source of energy. But there are problems disposing of the **radioactive waste** left by the fission process, and accidents like that in Chernobyl, USSR, in 1986, are always a possibility. Safer alternatives, such as solar, wave and wind power are now available.

The international sign for radiation danger is:

Conservation of fossil fuels

A lot of energy is produced by burning fossil fuels, but much of this energy is wasted. Conserving energy means being more efficient – being smarter.

There are six main ways to conserve fossil fuels:

- Increase the use of liquefied petroleum gas (LPG).
- Make liquid fuel from natural gas or coal.
- Produce alcohol fuels from renewable plant material.
- Use fuels more effectively and reduce wastage.
- Use more of the renewable sources.

The more we conserve energy the more time we have to improve renewable energies.

Renewable energy resources

Some of our energy sources will last for billions of years because they are being replaced all the time. These energies are renewable, clean and safe. However, some of these technologies do damage plant and animal habitats.

Wind power

Wind is produced when the land heats up and patches of warm air rise upward and cooler air rushes in to replace it. Air movement creates our weather.

For thousands of years people have used the energy of the wind for such jobs as grinding wheat and irrigation. About 100 years ago, windmills were installed in remote areas of Australia to help with such tasks.

Australian windmill

Modern wind generator

Wind energy today ranks as one of the most promising of the renewable energy technologies for generating electricity. It is anticipated that in the next few years the installation of wind power systems will increase significantly.

Australia's first commercial **wind farm** has been operating at Esperance in Western Australia for six years.

Wave power

The movement of waves, caused by the wind, is a form of water power. Water moves up and down inside a tube, which drives air through the turbine. In Bass Strait, the people of King Island use wave power.

Hydro-electric power

Hydro-electric power stations generate electricity using water trapped behind dams. The water stores potential energy because it is raised. This energy can be transferred to kinetic energy by letting the water flow downhill. The running water turns a turbine, which generates electricity.

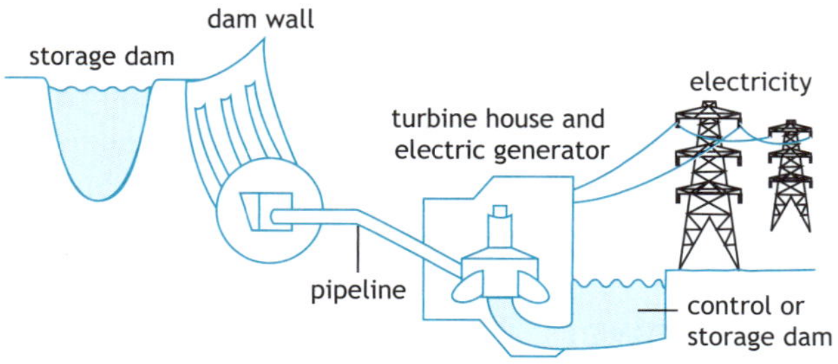

Hydro-electric station

Tidal Power

In coastal areas the level of the sea rises twice every day. **Tides** occur because the gravity of the sun, moon and planets is strong enough to move the water towards them. Water can be trapped behind a dam at high tide and used to drive a turbine when it is let out.

Geothermal power

Deep down in the ground the rocks are very hot due to **radioactive substances**. Hot rocks can be used to change water into steam, which can be used to generate electricity.

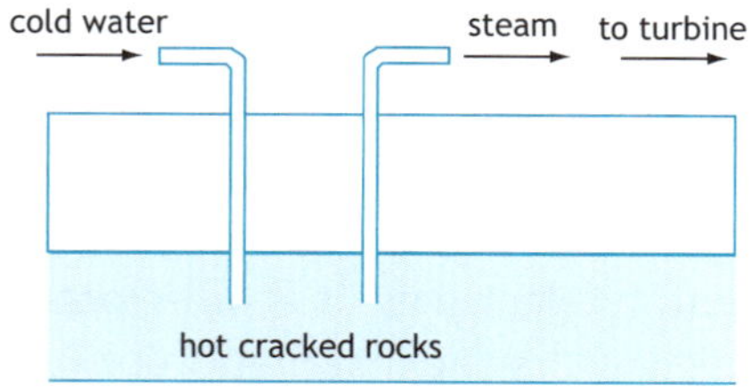

In some places the crust of the earth is so thin that the heat can escape through cracks, for example **hot springs** and **geysers**.

Solar power

Every day lots of energy, in the form of heat and light, reaches us from the sun. **Solar energy** is transformed into electrical energy in solar cells and solar panels.

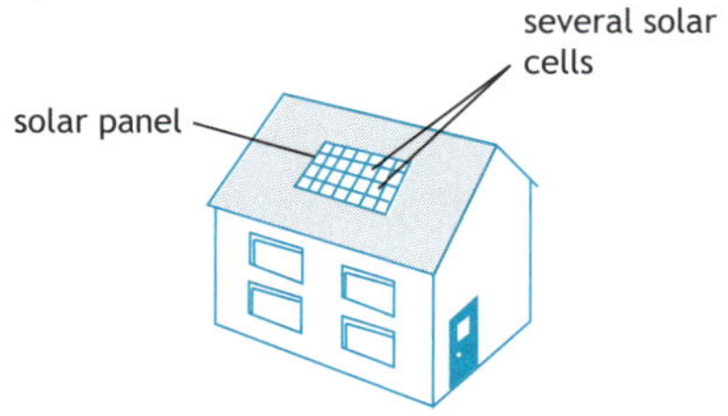

A solar panel

Solar energy is used to heat water, cool houses, power cars and generate electricity. Every year Australia hosts a solar car race from Darwin to Adelaide.

Solar car race

Biomass power

Biomass is any matter derived from **biological sources**, such as plants, animals, and any of their wastes. Using waste as an energy source is not a new idea.

Cow manure is a source for producing biomass energy

Biogas, or methane, is produced commercially using waste and bacteria inside an airtight container called a digester. Some rubbish dumps are now using digesters to generate methane. The methane is used as a fuel in furnaces. The waste from digesters is good fertiliser.

Power used in sugar mills is generated using **bagasse** (sugar cane processing waste) as a fuel.

Australia's energy resources

Australia has abundant stores of energy resources.

- Our proved, economic oil reserves are limited.
- We have very large coal deposits, equal to 1000 years' supply, at our current domestic rate of use. Australia has become the world's largest exporter of coal.
- Our gas reserves will continue for some years (maybe 60 years) at our current rate of use.

However, the level of energy use has become an important issue due to **environmental problems**. Scientists agree though, that a combination of fossil fuel conservation and the development of renewable resources is a good start towards solving these problems. Australia has some renewable technologies, such as solar and wind power, but there is not really one solution to our energy problems. Many solutions will have to be combined and we may need to move away from large power stations to local 'energy units', making use of renewable energy sources appropriate for the area.

The question is, can we provide all the energy our society needs without damaging the capacity of the Earth's natural systems? No-one knows. Research into how to provide our energy needs with the least environmental impact continues.

Glossary

Fossil fuels – Coal, oil and natural gas are fuels formed from the remains of trees and other organic matter long ago. They are usually buried deep underground.

Non-renewable – Energy sources that will eventually run out.

Renewable – Energy sources that won't run out because they are continually being replaced.

Science skills

Question 1

Study the pie graph below.

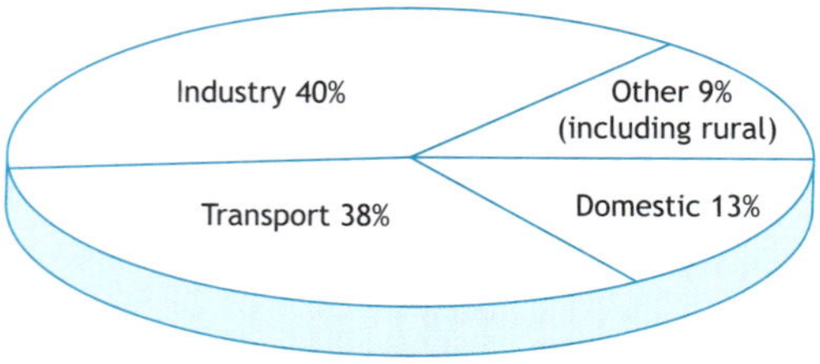

a What do Australians use the most energy for?

---

b What percentage of energy does transport use?

Question 2

The graph below displays the sources of energy in Australia.

a Which source of energy do we use the most?

---

b Which do we use the least?

---

c What is coal mostly used for in Australia?

---

d What are oil products mostly used for in Australia?

---

e Which of these sources are renewable?

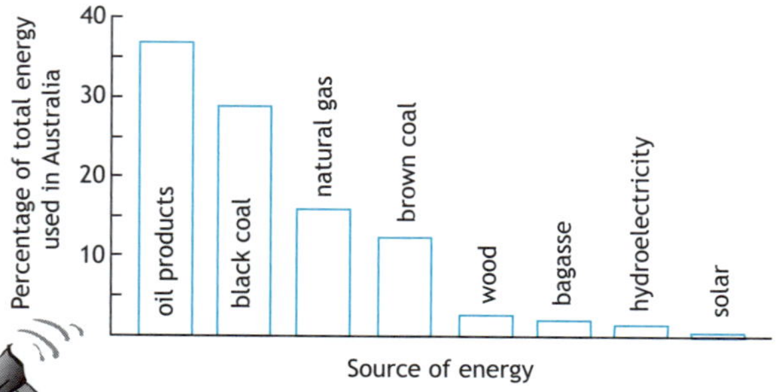

Question 3

The following diagram shows how a power station works. Label the components with the following terms:

generator, wire, boiler, switch, turbine, coal, television

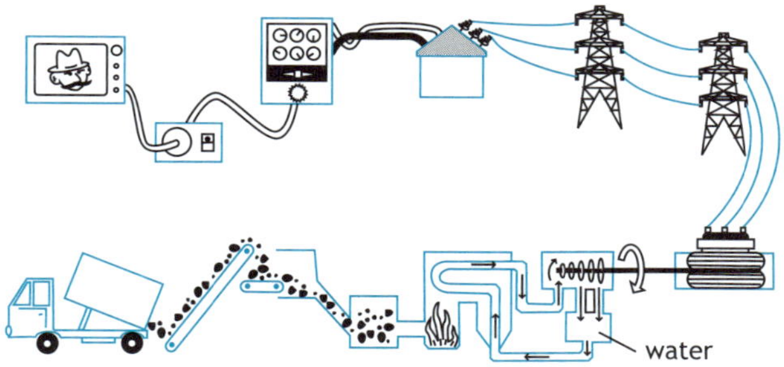

Question 4

The graph below shows the use of coal in Australia.

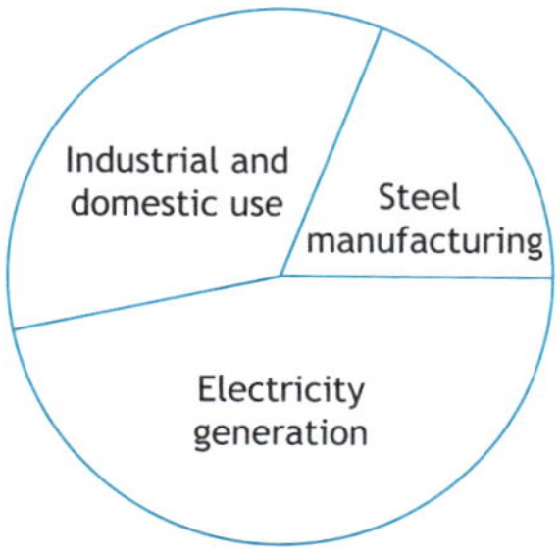

a For what purpose is coal used the most?

---

b Draw a bar graph to display this data.

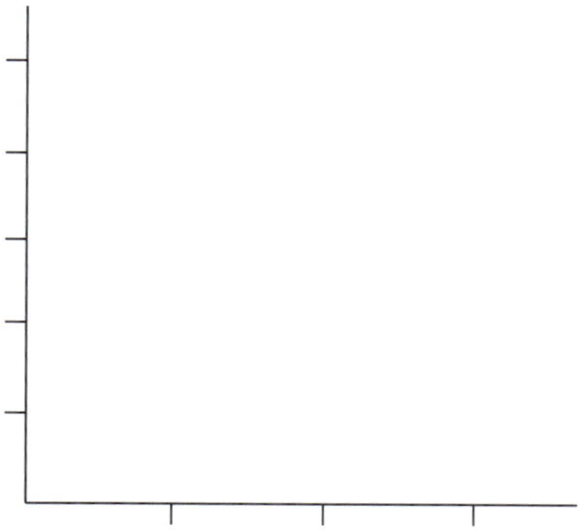

Question 5

The following diagram outlines the use in Australia of 100 barrels of oil.

a List the uses in order, from the largest number of barrels to the smallest.

b What percentage of oil is used for 'other purposes'?

c What is the largest percentage of oil used for?

d What is kerosene used for?

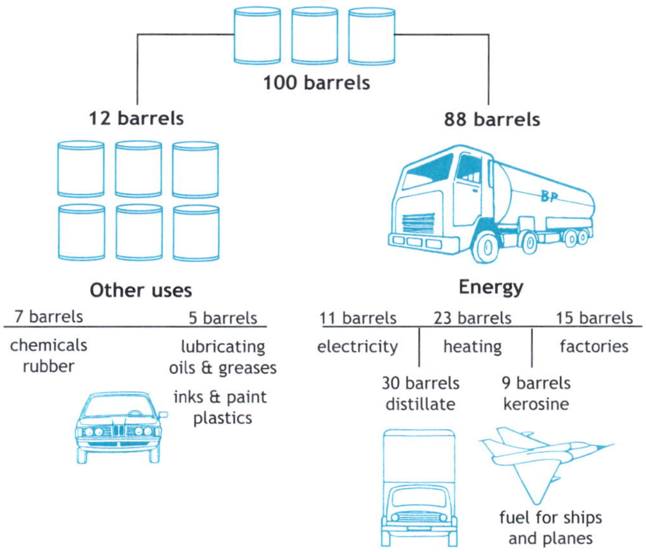

Question 6

The table in the next column shows the approximate daily per capita consumption of energy in various categories for five different lifestyles.

Lifestyle	Energy consumption by category (10^6 joules/day)			
	Food consumption	Homes and offices	Agriculture and industry	Transport
Hunters	12	8	–	–
Primitive agricultural society	17	17	17	–
Advanced agricultural society	25	50	29	4
Industrial society	29	134	100	10
Technology society	42	277	382	265

a Which category of energy use does lifestyle influence:

i the most?

ii the least?

b Assuming that hunter-gatherers in the past did not suffer from hunger, how do you explain the fact that people in modern technological societies consume more energy in their food than hunter-gatherers?

Question 7

Look at the following pictures and note examples of energy wastage in each picture.

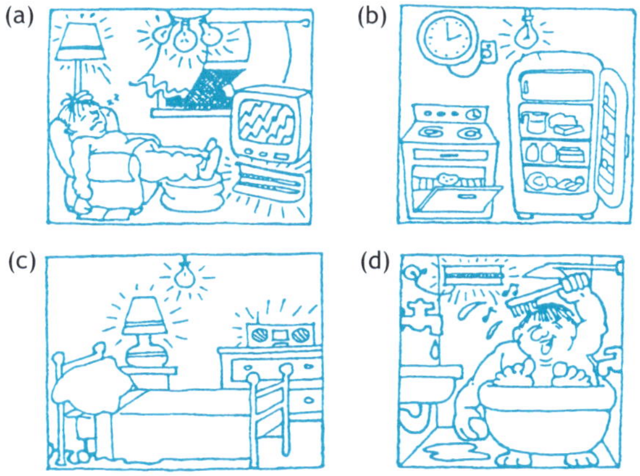

Question 8

The diagrams below represent the major environmental problems we face today in our pursuit of energy supply.

Match them with the correct description below.

- Radioactivity from nuclear fuel.
- Land degradation and other impacts, such as damage to plant and animal habitats from mining, dams and waste storage.
- Air pollution, both indoor and outdoor, from burning fossil fuels.
- Water pollution from waste, waste storage and accidental spills.
- The greenhouse effect.

(a)

(b)

(c)

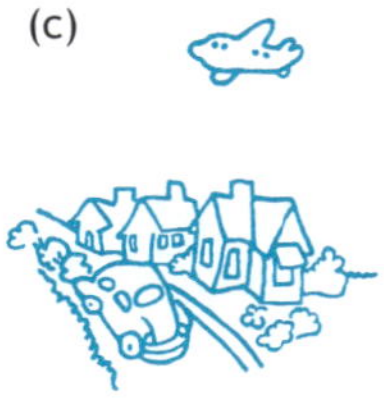

(d)

(e)

Question 9

Some students set up the experiment below by cutting a paper spiral and suspending it over a light bulb.

What form of energy is being created and used to spin the spiral?

____

Question 10

The diagram below describes the process of producing methane (biogas).

Fill in the boxes with the correct term listed below:

methane storage tank, electricity production, digester, organic matter collection, burning biogas, fertiliser

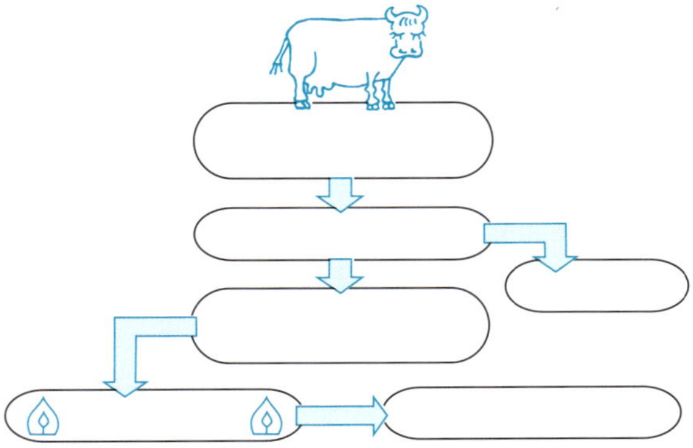

Question 11

A student decided to build herself a 'sunmobile'. She purchased the following:

Balsa wood, wheels, on-off switch, wire, glue, and sticky tape.

However, she forgot two very important parts.

a Look at her plan and write the down the two things she forgot.

____

____

b Explain the function of the two missing parts.

____

____

____

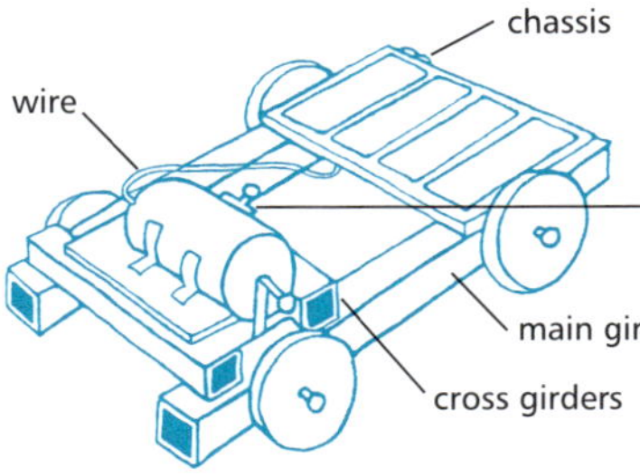

Question 12

The following table lists several renewable energy resources. Give one advantage and one disadvantage for each. Do not repeat either an advantage of disadvantage.

Source	Advantage	Disadvantage
solar		
tidal		
wind		
hydro		

Question 13

Australia has plenty of sunshine from the north in summer and cold winds from the south in winter. Builders are now positioning homes to make maximum use of sunshine in winter, while minimising heat in summer.

The diagram below shows an energy-efficient house. Study its construction and position then list at least five energy-saving concepts.

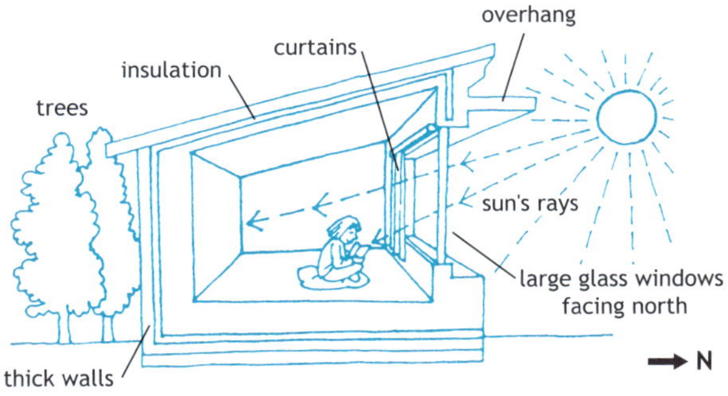

Question 14

The diagram below shows the heat loss and gain in an ordinary house.

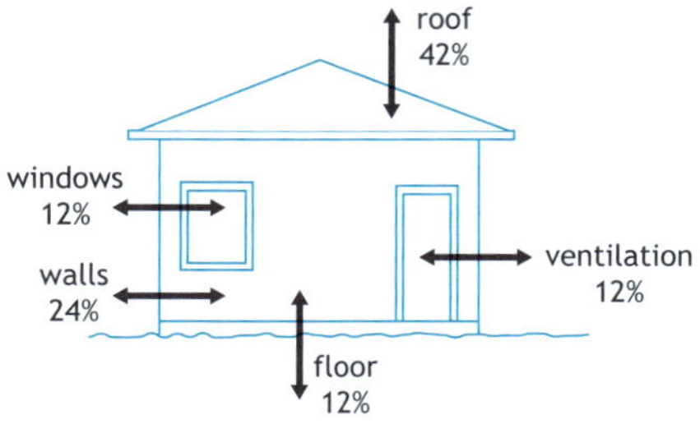

a Through which area is most heat lost or gained?

b What could be done to reduce this problem?

c What can be done to windows to prevent heat loss?

Question 15

Country	Units per head of population
USA	100
Germany	50
France	40
Sweden	65
Japan	37
Australia	55

a Draw a column graph to display the data above, which lists the energy consumption per head of population for several western countries.

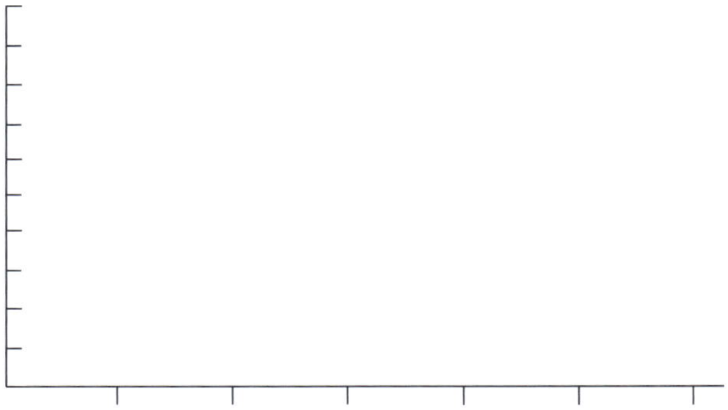

b Do you think that countries such as India, Nigeria, Ghana and Thailand would have energy consumption similar to the countries in the graph?

Question 16

Look carefully at the chart below. It is a forecast of energy supply and demand in the future.

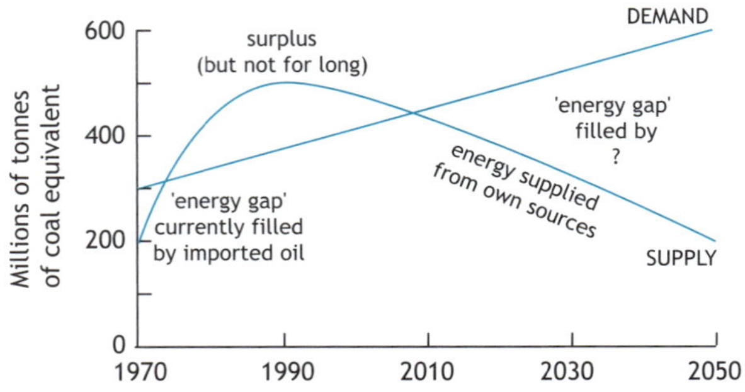

Projected UK energy demands and supply

a What is meant by 'energy demand'?

b What is meant by 'energy supply'?

c What is the surplus in the middle of the graph?

d According to this graph, when does the surplus end?

e What was the energy 'gap' on the left-hand side?

f How do you think the future energy gap on the right-hand side will be filled?

Question 1

Energy is stored in coal. This energy originally came from: a the sun.

b the centre of the Earth.

c chemicals.

d electricity of thunderstorms.

Question 2

a What does the 'conservation of energy' mean?

b What are some ways of conserving our energy?

c Why is it a good idea to conserve energy?

Question 3

Which one of the following is not considered a renewable energy source?

a tidal energy

b hydro-electric energy

c coal-burning power station

d solar energy

Question 4

In Alaska a few years ago, a huge tanker called the *Exxon Valdez* spilled millions of litres of crude oil into Prince William Sound.

What are the environmental consequences of an accident like this?

Question 5

Burning petroleum products produces poisonous gases.

a List two gases produced.

b What name do we give to air pollution produced by the exhaust of cars?

c What are two other environmental problems caused by this type of air pollution?

Question 6

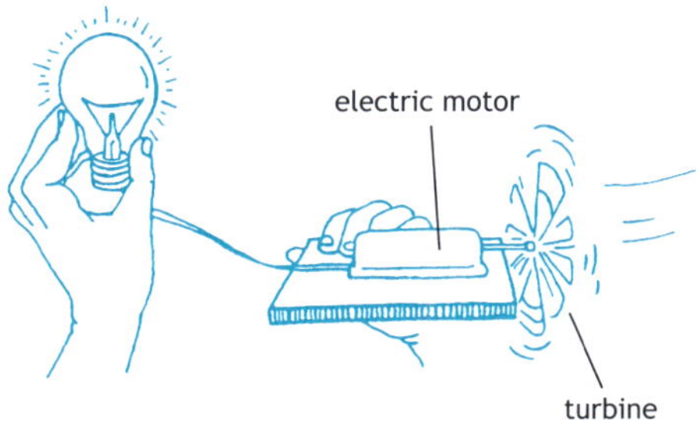

a What type of energy is powering this light bulb?

b Is this a renewable energy or not?

c What other forms of energy could you use to turn the light on?

Question 7

Name two natural features that result from heat escaping from the Earth through cracks in its crust.

Question 8

Fill in the missing words:

Hydro-electricity comes from falling ___________, while ___________ and ___________ power comes from the oceans. All of these need moving ___________ to make ___________ . The movement of water spins ___________, which generate ___________.

Question 9

The list below gives some sources of energy:

natural gas   oil   coal   nuclear   hydro

wind   wave   solar   geothermal   tidal

a Name two of these sources that are used to generate electricity on a large scale.

b When electricity is generated in a power station, much of the input energy is wasted. Give one reason why electrical energy is so useful that we accept this energy loss.

c Coal is a fossil fuel.

i Name one other fossil fuel in the list above.

ii Give one disadvantage of using fossil fuel.

d Solar energy is a renewable source of energy.

i Name one other renewable source of energy in the list above.

ii Explain what is meant by the term renewable energy.

iii Why would solar energy not be used on a large scale in the United Kingdom?

e What is geothermal energy?

Question 10

Arrange into the correct order the following sentences about how coal, oil and gas are formed.

a Trees store this energy as they grow.

b Dead trees fall onto swamps.

c The wood gradually turns into coal, gas or oil.

d Energy is transferred to trees by sunlight.

e The dead trees are buried under layers of mud.

Question 11

Explain how energy from the sun:

a produces wind.

b produces waves.

c keeps filling up the water behind dams.

Question 12

The following diagrams show several energy sources.

a Which two energy sources depend on radioactive substances?

b Which energy source depends mainly on gravity and the moon?

c Which sources depend on the flow of water?

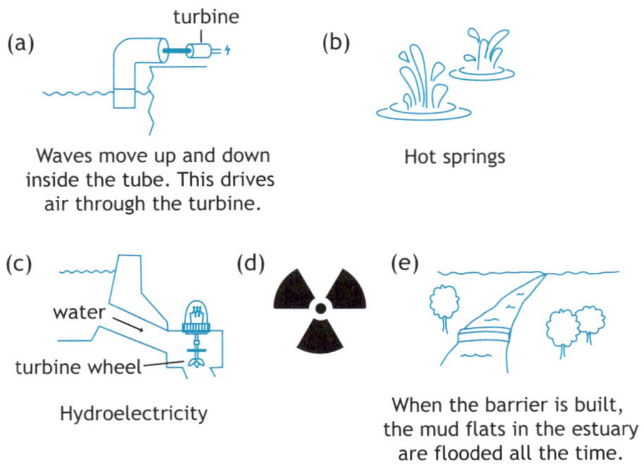

(a) Waves move up and down inside the tube. This drives air through the turbine.

(b) Hot springs

(c) Hydroelectricity

(e) When the barrier is built, the mud flats in the estuary are flooded all the time.

Question 13

a Oil is a non-renewable energy resource. Why?

b Most of the oil we use is burned as fuel. What is its main purpose in Australia?

c What else can oil be used for?

Question 14

The following is a list of energy sources. Write 'yes' against the ones that are renewable or not likely to run out for an extremely long time.

Energy source	Is it renewable?
wind	
coal	
solar	
nuclear	
biogas	
natural gas	
tidal	
geothermal	
oil	
waves	

Question 15

What three things do we need to burn fossil fuels?

____

____

____

Question 16

Scientists believe that the energy demand of contemporary society will be met by:

a ______________________________ and

b ______________________________

Question 17

At a coal-burning power station, coal is stored then crushed before it is fed into a furnace where it is burnt. Heat from the burning coal is used to change water into steam. Steam drives a turbine attached to a generator, which produces electricity.

Which of the following summarises this process?

a crushers – storage – turbine – burning – generator – electricity

b storage – crushers – burning – generator – turbine – electricity

c storage – crushers – burning – turbine – generator – electricity

d crushers – storage – burning – turbine – generator – electricity

Question 18

Use the graph of energy usage to decide which of these statements in incorrect.

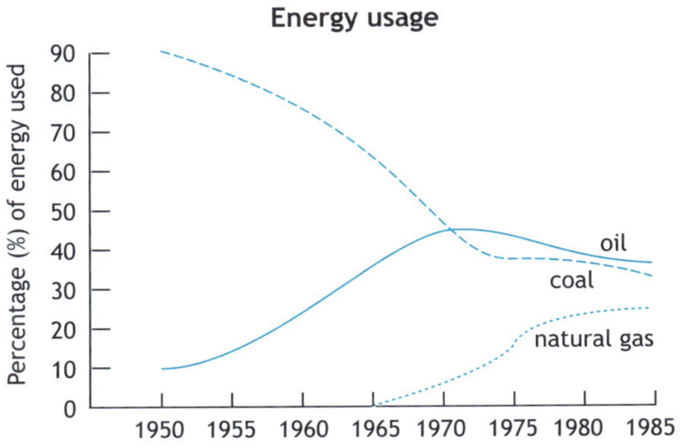

a Less oil was used in 1985 than in 1970.

b As the use of oil decreased the use of coal increased.

c We have always used more natural gas than oil.

d In 1970 the percentage of energy from coal began to increase.

Question 19

Many major countries have agreed to reduce their atmospheric carbon dioxide levels.

The amount of carbon dioxide in the air could be reduced by:

a burning oil instead of coal.

b clearing the trees from more of the countryside.

c planting more trees.

d keeping more farm animals.

Question 20

Use the data in the table to answer the following questions.

Biomass	Energy potential (rating of 1-5)
household waste	1
forest waste	2
sawmill waste	3
bagasse	4
cereal straw	5

a What biomass has the most energy potential?

____

b Which of the biomass sources would be the easiest to collect to power a city biogas plant?

____

c From which industry is bagasse a waste product?

____

CHAPTER 8

NUCLEAR ENERGY

Goin' fission is confusion

Fossil fuels provide 80% of the energy requirements of the modern world. To provide for the future we need to use more sustainable or renewable alternative energy sources. Some people believe that the answer is nuclear energy.

Nuclear energy is the energy that binds together the **neutrons** and **protons** in the nucleus of an **atom**. A change in the nucleus causes energy to be released. The energy from the sun and stars comes from nuclear reactions. Scientists can produce nuclear energy from some atoms, for example uranium, in a nuclear power station or a nuclear weapon.

In 1919 Ernest Rutherford, a scientist at Cambridge University, bombed atoms of nitrogen gas with alpha particles and changed them into oxygen and hydrogen atoms. These early experiments led to the splitting of the atom.

Nuclear fission and nuclear fusion

Fission and fusion are both forms of nuclear reaction.

- A **fission reaction** involves the splitting of a large nucleus, like uranium, by bombarding it with neutrons. Nuclear fission is used in nuclear power stations and atomic bombs. Nuclear fission does produce clean energy but it also produces dangerous wastes, which must be stored safely for a long time.
- **Nuclear fusion** is the fusing together of tiny nuclei, such as deuterium (a special form of hydrogen). This reaction is the most powerful source of energy known, but involves a minimum of dangerous waste. Fusion is used in hydrogen bombs but has not been successfully developed for power stations.

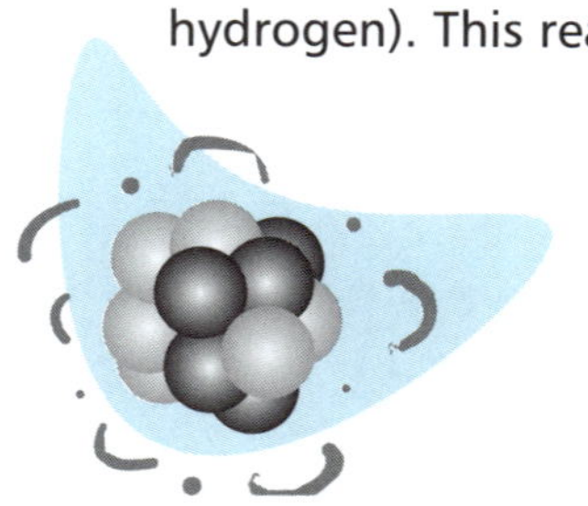

The sun's energy is produced by hydrogen atoms combining to form helium in a fusion reaction.

Chain reaction

A chain reaction is a series of events, one triggering another. When a neutron strikes a uranium atom, the atom breaks up. It produces three 'free' neutrons, which in turn strike other atoms and so on. This chain reaction produces heat energy.

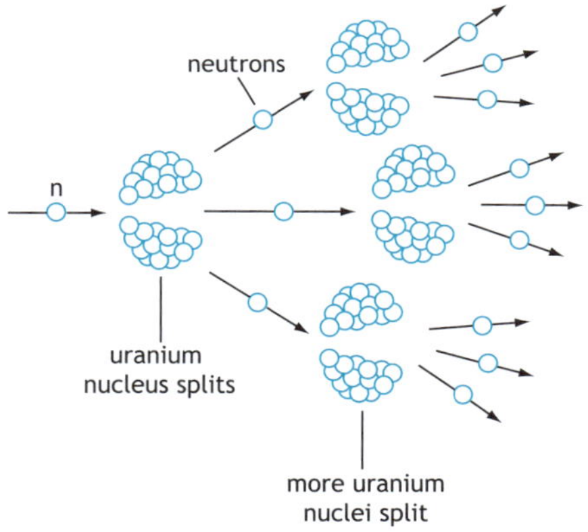

A nuclear chain reaction

Nuclear reactors

About 400 nuclear power plants already generate 20% of the world's electricity. However, a nuclear power plant only lasts about 40 years and they are difficult to dismantle because of the radioactivity. The problem of disposing of the radioactive waste from nuclear power stations has yet to be solved and nuclear accidents can be catastrophic in their consequences.

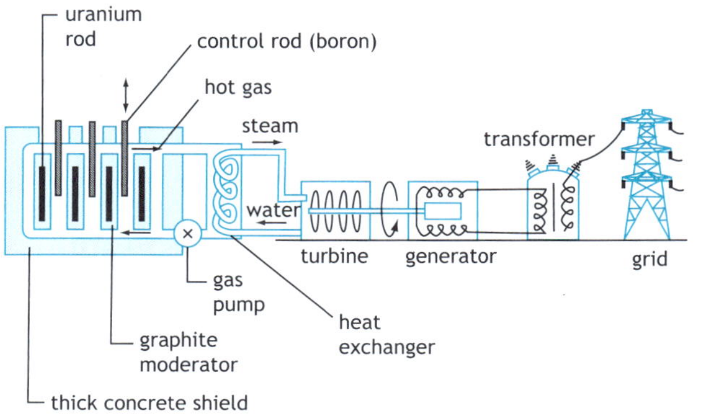

Gas-cooled reactor

nuclear energy ⇨ heat in coolant ⇨ potential energy of steam ⇨ kinetic energy of turbine ⇨ electrical energy ⇨

A nuclear reactor

- Inside a nuclear reactor, uranium fuel is sealed inside **rods**. The rods are arranged in groups called **fuel elements**. The fission chain reaction makes these rods produce heat. A liquid (water or a molten metal) or a gas flows around the rods to cool them. This is called a **coolant**.
- Special **control rods**, made of a substance which absorbs neutrons, can be raised or lowered between the fuel elements. When lowered these rods stop the chain reaction.
- To slow the neutrons down, and as an extra precaution against neutrons escaping, the elements are surrounded by a **moderator** (water can be used). The moderator deflects neutrons back into the reaction.
- To contain the **dangerous radioactivity** produced by the chain reaction, the reactor assembly is shielded by thick walls of steel and/or concrete.

Making electricity

A **nuclear power plant** works in the same way as an ordinary coal-burning power station, by heating water to make steam to drive turbines. The heat to make the steam comes from the enormous energy released by a chain reaction inside the plant's reactor.

Just like coal and oil, uranium is a non-renewable resource and will eventually run out. So scientists have invented the 'Fast Breeder Reactor' which uses uranium to produce **plutonium**, which is also used as fuel. This reactor can produce 50 times more energy than a normal reactor. However:

- the chance of an explosion is much higher.
- plutonium is extremely dangerous because it has a long half-life and is therefore very radioactive and toxic.
- there is no agreement internationally as to what to do with the waste.
- nuclear bombs can be made from plutonium.

Nuclear weapons

There has been some reduction in nuclear weapons by the super powers in recent years; however, if all the stockpiled nuclear weapons were used, much of the biosphere and probably all human life and culture would be destroyed.

• The atomic bomb

In an atomic bomb, uranium is bombarded with neutrons, which begins an uncontrolled chain reaction. A massive amount of energy is produced causing an explosion. Heat and radiation are produced at a rapid rate. The first atomic bombs were dropped on the Japanese cities of Hiroshima and Nagasaki during World War II. These bombs destroyed both cities, killed 200 000 people and ended the war.

• The hydrogen bomb

A hydrogen bomb gets its energy from nuclear fusion. An atomic bomb must be exploded first to generate the enormous heat needed to set off an H-bomb.

• Nuclear submarines

A nuclear submarine is driven by electricity generated by a nuclear power reactor. Nuclear submarines do not need to be refuelled.

What happened at Chernobyl

In April 1986 there was an accident at the Chernobyl nuclear power plant in Russia. A reactor exploded and radioactive gases spilled out into the air. Large areas of countryside were contaminated by the radioactivity, which affected plants and people. The reactor was too dangerous to repair and was sealed in concrete – possibly forever.

Radioactivity

Some elements, called **radioisotopes**, are naturally unstable and emit radioactive rays which can affect their surroundings. There are three main types of radiation:

1. Gamma rays – these are short-wave electromagnetic radiation and can be stopped by 1 cm of lead or 1 m concrete. These waves are the most penetrating and the most dangerous.
2. Beta rays – these are fast moving streams of electrons and can be stopped by 3 mm of aluminium.
3. Alpha rays – these are positively charged particles moving at high speed and can be stopped with paper or clothing.

Uranium is one substance that naturally emits radioactive particles. This process is called radioactive decay. The average time taken for half the atoms present at any time to decay is called the **half-life** of the element. Knowing an element's half-life can be useful. For example, carbon-14 can be used to tell the age of carboncontaining fossils or old furniture. Carbon-14 has a half-life of 6000 years.

Working out the age of fossils

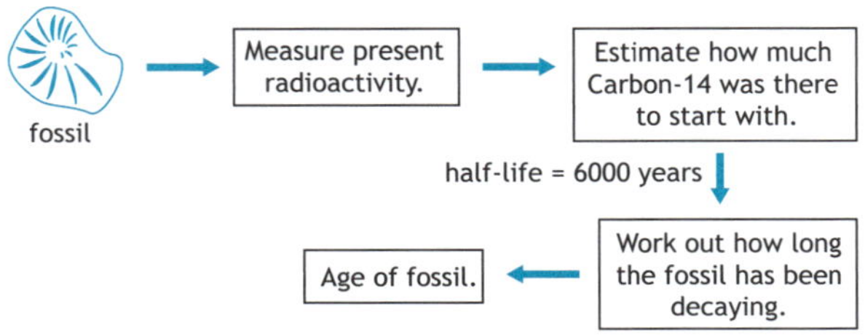

Radioactivity is dangerus because of the energy involved. Some forms of radiation can last for thousands of years. Nobody knows how to make exessive radiation safe and scientists are careful not to let radiation escape.

However, small amounts are natural and are around us all the time. This is called **background radiation**.

Some radioactive substances can be useful.

- In medicine, radioactive tracers or radioisotopes can be used in patients to locate organs not functioning properly. Pacemakers, which keep the heart beating regularly, are powered by a tiny nuclear power unit.

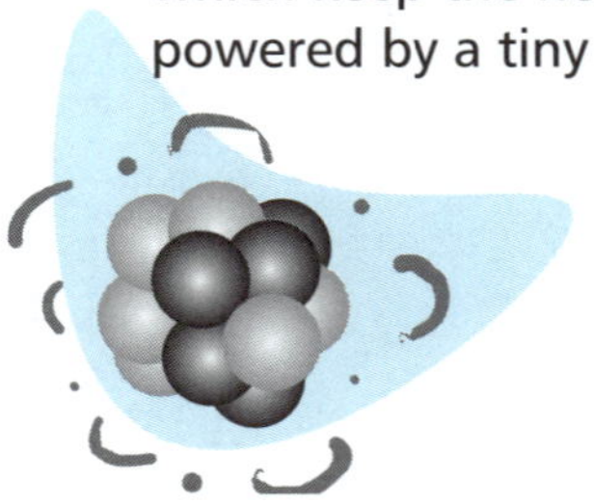

- Scientists use tracers to study the path of fertilisers through a plant.
- Industry uses radioisotopes to test oil for piston wear in cars.
- Some satellites that orbit the Earth are nuclear powered. Future rockets will use nuclear power, as they will travel too far to use solar power and will not be able to refuel for years.

Radiation levels can be measured in 'counts/minute' using an instrument called a **Geiger counter**. Geologists may also use this instrument when searching for uranium.

Nuclear power and the environment

Advantages of nuclear power

- It produces little atmospheric emission. It does not produce carbon dioxide so it does not contribute to the greenhouse effect, acid rain or local pollution from particles.
- It does not need coal. Countries without cheap coal supplies are very keen on nuclear power.

Disadvantages of nuclear power

- *The transport and disposal of radioactive material.* The radiation from spent fuel rods is very damaging to all organisms. It affects the normal working of the cells in the body and can lead to cancer and leukemia. Plutonium has a half-life of 24 000 years. Nobody knows where to store such materials so that they will be safe for this long. Some scientists are trying to bury it deep under ground and others are using a synthetic rock called Synroc.
- *The removal of the old power plant.* Nuclear power plants have a limited life and many of the world's plants are now close to being dismantled. This is expensive and nobody knows where to store the most radioactive parts.
- The cost of constructing a safe nuclear power plant is much more than a coal-fired plant, but they are cheaper to operate. Permanent storage of high-level radioactive waste is very expensive.
- Nuclear power plants are potentially very dangerous and if problems do arise the result can be disastrous. Newer reactors are considered to be much safer than the older ones but public perceptions remain a major disadvantage.

Nuclear power and Australia

There is no nuclear power generation in Australia. A small nuclear reactor operates at Lucas Heights near Sydney, producing small quantities of radioactive substances for medical and scientific uses. However, Australia has large reserves of uranium. Mining this uranium leaves some low-level radioactive waste but this does not present the problem of high-level waste from reactors. Australia does have some high-level waste from the Lucas Heights reactor and ways of dealing with it are currently under discussion.

Glossary

Control rods – Absorb the neutrons in a nuclear reactor.

Coolant – Water, gas or molten metal used to cool the fuel rods in a nuclear reactor.

Fission – The splitting of larger atoms into smaller atoms.

Fuel rods – Rods in a nuclear reactor where uranium undergoes a chain reaction.

Fusion – Involves making larger atoms from smaller atoms.

Geiger counter – Instrument used to measure levels of radioactivity.

Half-life – The average time it takes for half the atoms present in a substance to decay.

Moderator – A substance used in most nuclear reactors to slow neutrons down and prevents any escaping.

Neutrons – Electrically neutral particles in an atom's nucleus.

Nuclear power – The generation of electric power by nuclear fission or fusion reactions.

Nuclear power reactor – A reactor used to generate electricity.

Nucleus – The dense central core of an atom consisting of protons and neutrons.

Proton – Large positively charged particles in the nucleus of an atom.

Radiation – The spread of electromagnetic radiation and other subatomic particles.

Radioactivity – The spontaneous break-up of certain nuclei, resulting in the emission of alpha rays, beta rays and gamma rays.

Science skills

Question 12

The half-life of carbon-14 is 6000 years.

a In the diagram below, how many half-lives have passed since the start?

b If this was a fossil, how old would it be?

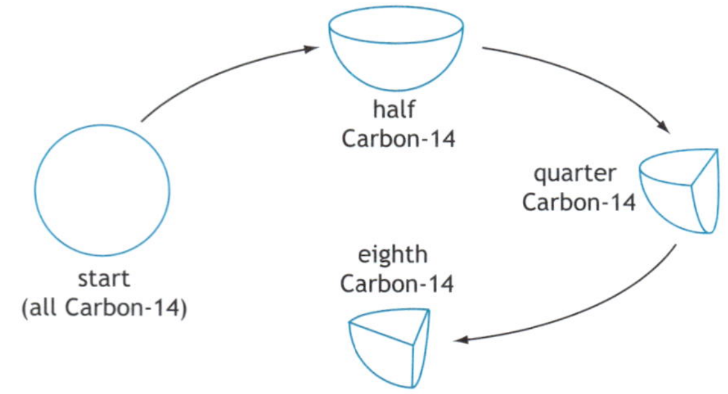

c The substance strontium has a half-life of 28 years. It emits beta particles, which can be absorbed into human bone. How many years must pass before its activity falls to one-sixteenth of its original value?

Question 2

The following table gives values for the activity of a sample of a radioactive substance at different times.

Time (minutes)	Activity (counts/second)
0	96
10	78
20	62
30	54
40	40
50	32
60	26
70	21
80	15
90	14

a Plot a graph to show how the activity of the sample changes with time.

b State a value for the half-life of the substance.

Question 3

A group of students were set the task of measuring the half-life of a radioactive substance that emits beta particles only. They measured the count rate every 20 minutes and recorded their readings as shown below.

Counts/minute	Time (minutes)
330	0
321	20
165	40
120	60
90	80
71	100
57	120

They discovered a count rate of 30/min was obtained even when there was no radiation source near.

a What was the cause of the count rate of 30/min?

b Complete this table showing how the count rate due to the beta source only (does not include background) varies with time.

Counts/minute	Time (minutes)
	0
	20
	40
	60
	80
	100
	120

c Plot a graph of the beta count rate against time and use it to find the half-life of the source.

Question 4

Draw a flow chart representing the production of electricity from a nuclear reactor using the terms listed below:

steam, dynamo, nuclear reactor, electric energy, heat, moving turbine, nuclear energy.

Question 5

Nuclear energy is the energy that holds atoms together. Splitting the atom releases this power.

Label the diagram below with the following terms:

neutron, neutrons, atom, energy

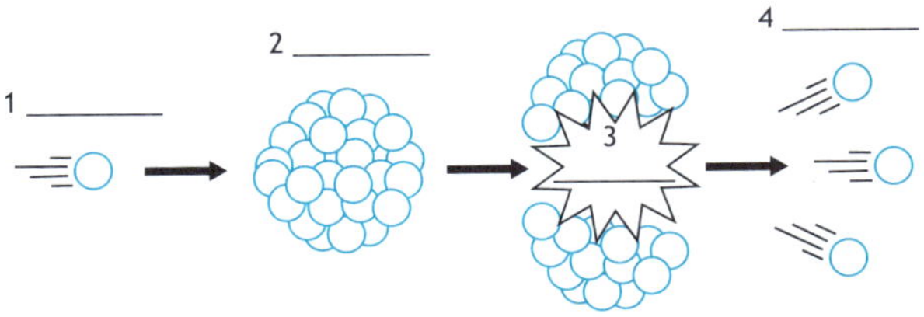

Neutron collides with the centre or 'nucleus' of an atom

Question 6

Label the parts of the nuclear power plant drawn below.

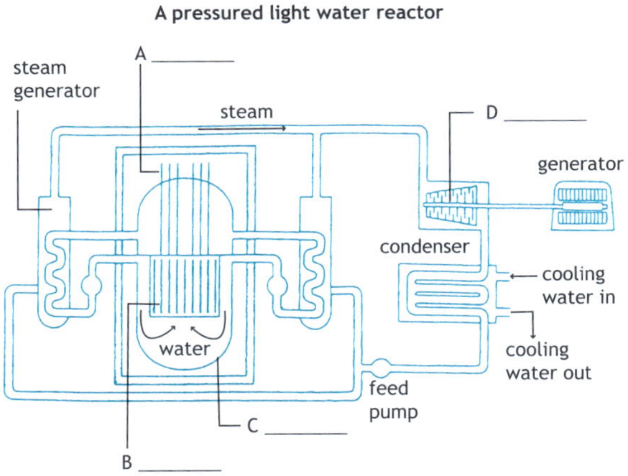

Question 7

Which of the following are advantages of nuclear power?

a Transport of nuclear waste.

b Produces little atmospheric emission.

c Cost of the disposal of radioactive waste.

d Cost of construction of the plant.

e The useful life of a power plant.

f Does not use coal.

Question 8

Match the following parts of a nuclear reactor with the correct description.

Fuel	Slows down the neutrons.
Shield	Control the chain reaction.
Coolant	Uranium.
Moderator	Protection from radiation.
Control rods	Carries heat away from reactor.

Question 9

What special name is given to the following reaction?

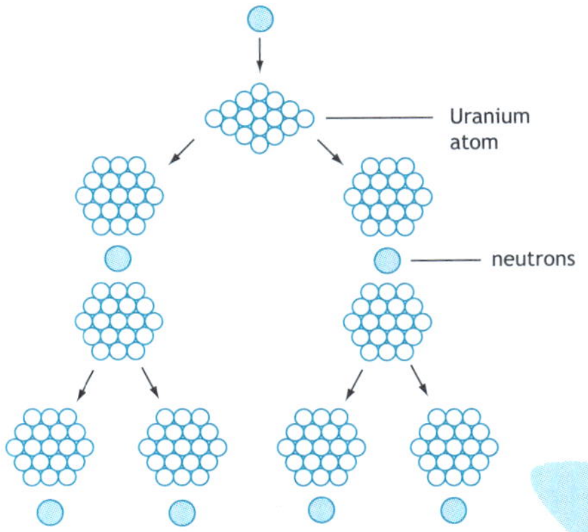

Question 10

The graph below shows a decay curve for a radioactive substance.

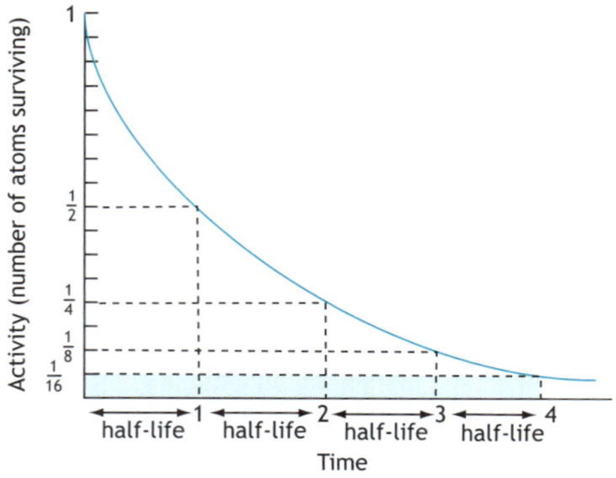

a How many atoms have decayed during one half-life?

b How many atoms have survived two half-lives?

c What fraction survives after four half-lives?

Question 11

Write the correct heading for each type of radiation.

Properties			
nature	electron	very short waves	positively charged
penetration	3 mm aluminium	reduced by lead	stopped by paper or skin
dangerous	yes	yes	yes

Question 12

In a rock sample it was found that for every 5 atoms of uranium when the rock was formed, 1 atom had decayed and 4 atoms had not.

a What percentage of the atoms has survived?

b Study the following half-life graph and work out how many years it has taken to reduce the activity to this level.

c How long will it take to decay 75% of the uranium?

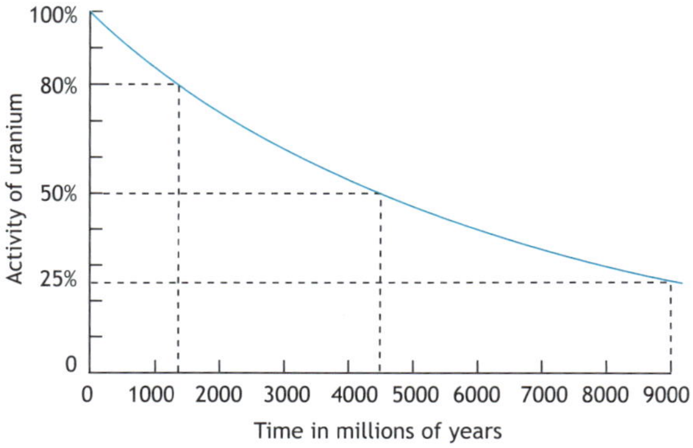

Question 13

A radioactive mixture is made up of element X, which gives a count rate of 160 per second and has a half-life of 2 hours, and element Y, which also gives a count rate of 160 per second but has a half-life of 1 hour. If the initial count rate of the mixture is 320 what will be its value after:

a 2 hours? ____________

b 4 hours? ____________

Question 14

Study the graph below and answer the questions.

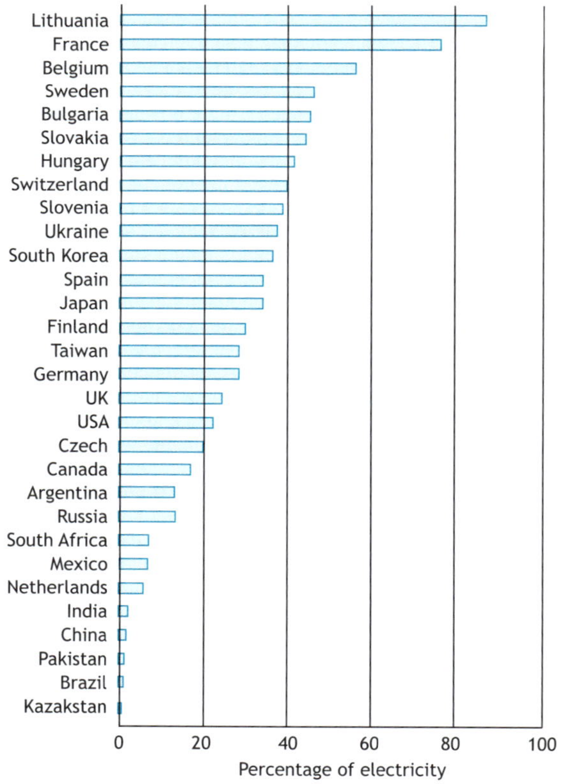

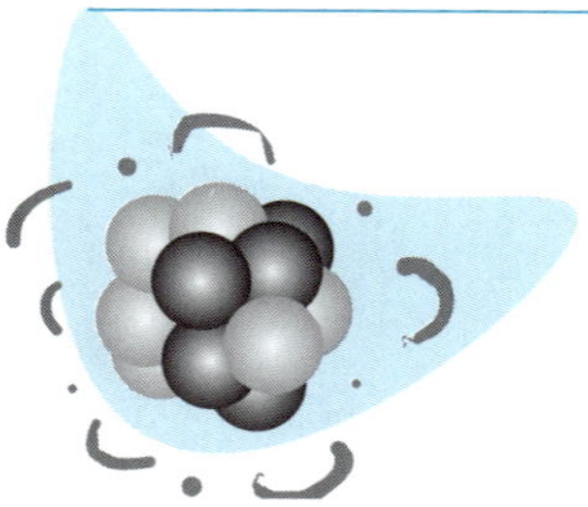

a Which country has the most electricity supplied by nuclear power?

b What percentage of Switzerland's electricity is supplied by nuclear power?

c Electricity supplied by nuclear power accounts for how much of France's power needs?

d On which continent is most of the nuclear power generated?

Question 15

The diagram below shows the three particles emitted by radioisotopes and substances that will stop them. Label the rays.

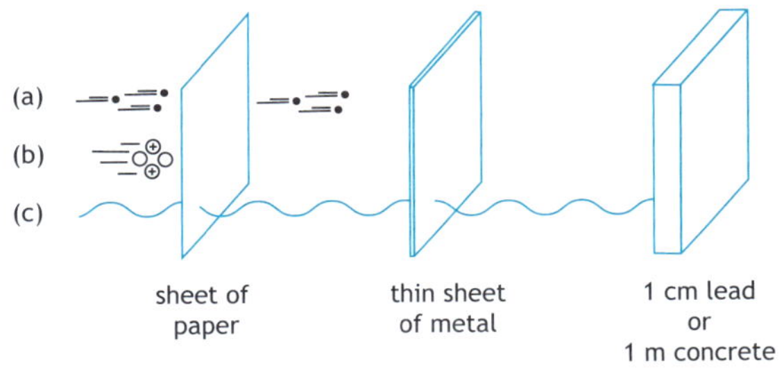

Question 16

When using radioactive tracers it is important to choose the right one. If the half-life is too short, the radioisotope may cease its activity before the experiment is complete. Because the radiation could be dangerous, you don't want the half-life to be longer than necessary.

For each problem below, check the table and then decide which radioisotope you would use. Give reasons for your answers.

Tracer	Type of radiation	Half-life
Polonium-210	alpha	138 days
Phosphorus-32	beta	14 days
Iridium-192	gamma	74 days
Technetium-99	gamma	6 hours

a Underground pipes sometimes spring leaks that are hard to find. It is expensive and impractical to dig up large sections in the hope of finding leaks.

b You are a hospital consultant specialising in the treatment of respiratory problems. You suspect that one of your patients has a blockage in an air passage in one of his lungs (X-rays are not an option).

c An agricultural chemist has developed a new herbicide that is intended to attack only broad leaved weeds and not be absorbed at harmful levels by grain crops.

d A doctor wants to assess the functioning of the pancreas in children with cystic fibrosis. This process takes several days.

Revision questions

Question 1

What is the name of the scientist who first conducted experiments that eventually led to the discovery of nuclear fission?

Question 2

What effect does radiation have on living organisms?

Question 3

Indonesia claims to need nuclear power to meet its rapidly increasing demand for electricity. It says its oil and natural gas supplies are reducing. The Indonesian government plans to build a plant on the island of Java, where 100 million people live. This area is also prone to earthquakes and volcanoes.

a Why are so many people very concerned about the building of this plant?

b Could this nuclear plant affect Australia?

Question 4

Choose the correct answer.

Control rods within nuclear reactors:

a absorb neutrons to keep fission at a steady rate.

b slow neutrons down to allow fission to take place.

c supply uranium for fission to take place.

d speed neutrons to bring about fission.

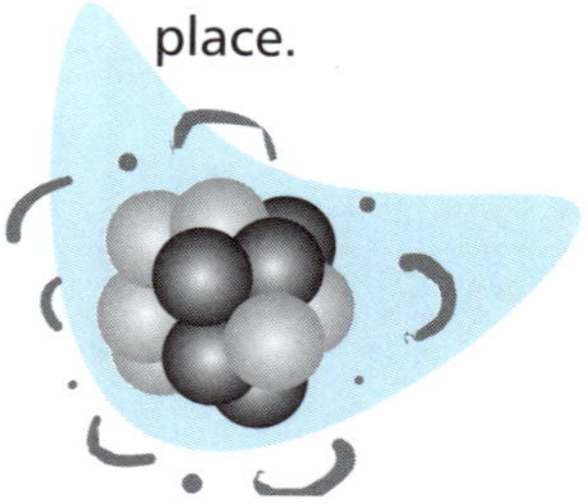

Question 5

Nuclear fusion is the most powerful type of nuclear reaction.

a It involves forcing very light atoms of ____________ together.

b Have we learned to produce electricity from fusion reactions?

c Where do fusion reactions occur naturally?

d Which nuclear weapon uses a fusion reaction?

e What type of nuclear reaction is needed to provide the heat to start the fusion reaction in the H-bomb?

f What is a fission reaction?

g Which nuclear weapons depend on fission reactions?

h Can we use nuclear fission to produce power?

Question 6

Fill in the correct words.

________ is the major problem with the nuclear power process. The fuel handling equipment becomes dangerous for __________ of years and the used fuel ______ release dangerous amounts of ____________.

Question 7

How does a breeder fission reactor differ from a nuclear reactor?

Question 8

a What instrument is used to measure radiation levels?

b What measurement does it record?

____

c What is 'background radiation'?

____

____

Question 9

Study the diagram below and then fill in the missing words.

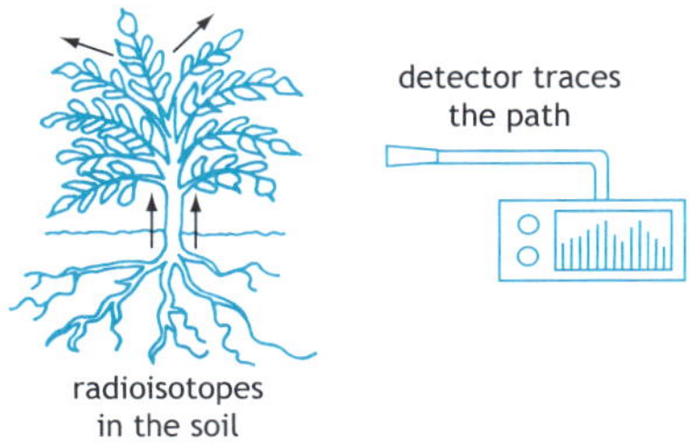

To investigate the ways that plants grow, chemicals containing ____________ are injected into the ______ around the plants. The plants ________ the radioisotope and then its path through the plant can be traced by a __________.

Question 10

Decide which of the following statements is true or false.

a Australia has no uranium. ______

b Nuclear power stations do not contribute to the greenhouse problem. ______

c Radioisotopes used in medicine are made in nuclear reactors. ______

d There have been no major accidents with nuclear reactors. ______

e There is a nuclear reactor near Sydney. ______

f The storage of spent fuel rods is no longer a problem. ______

Question 11

Fill in the missing words:

When a _________ atom splits it releases three _________. These neutrons may hit another uranium ________, splitting it as well. This in turn produces more ________ and ________, and so it goes on. This is called a ________ ________.

Question 12

A nuclear reactor producing energy by the fission chain reaction process requires a moderator. The function of the moderator is to:

a moderate the temperature of the reactor.

b split the atoms.

c regulate the number of fission producing neutrons.

d slow down neutrons without absorbing them.

Question 13

a Which of the following statements describes a fission reaction?

i Enormous power is released by splitting large atoms.

ii Enormous amounts of energy are produced when small atoms are forced together.

b Which process does the sun run on?

____

Question 14

a What is the main hazard of radiation?

____

b Why is the disposal of long half-life radioactive waste a problem?

____

____

c How could the accumulation of radioactive waste in the country possibly affect people in the city?

____

____

d Explain the statement 'Fusion reactors would be safer for society than fission reactors'.

____

____

Question 15

a Where does the energy for nuclear power come from?

____

b What are the three main types of natural radiation emitted by radioactive substances?

____

c How do we measure the rate at which radiations are given out?

____

d Can radioactive substances be useful as well as dangerous?

____

Test B

Question 1

The diagram below shows the path of cyclone Tracy over Darwin in 1974.

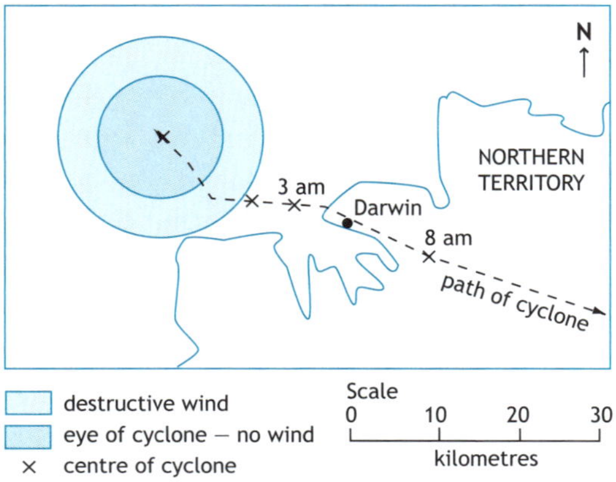

a Does a cyclone rotate clockwise or anticlockwise?

____

b From which direction was the wind coming in Darwin at 3 am?

____

c From which direction was the wind coming at 8 am?

____

Question 2

Label the diagram below with the following information:

isobars, air pressure (hectopascals), wind direction in Brisbane, the correct pressure system (high or low), the weather in Sydney

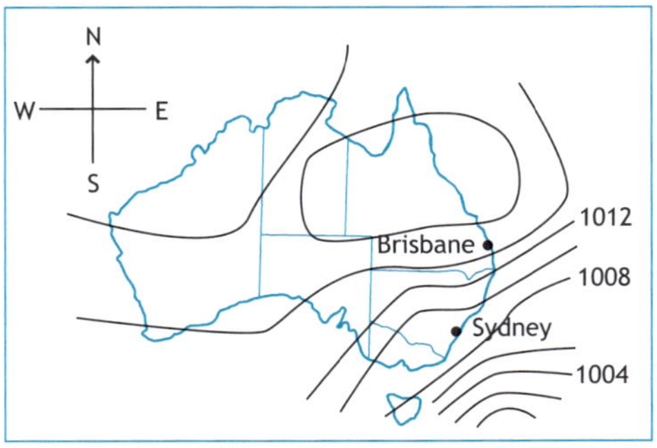

Question 3

Fill in the missing words.

__________ are lines on weather maps that join points of equal ______ __________. Air pressure is measured in ______________. Weather patterns in Australia move from ________ to ________.

Question 4

The following graph displays:

- monthly maximum and minimum temperatures
- average maximum and minimum temperatures
- the mean monthly temperature (average of lines 2 and 4).

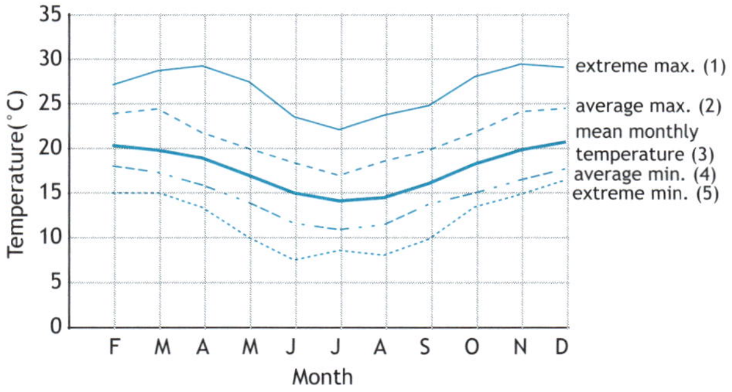

a What is the average minimum temperature for July?

____

b Which month has the lowest minimum temperature?

c What is the mean monthly temperature for November?

d What is the average maximum temperature for December?

Question 5

The following diagrams show a maximum/minimum thermometer.

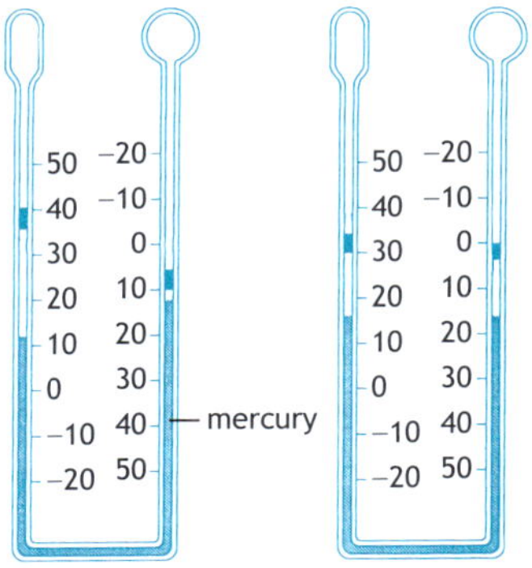

On day 1 the maximum temperature was 36 °C and the minimum was 10 °C. What was the maximum and minimum temperature on day 2?

Question 6

The following diagrams show how wind spirals around highs and lows in the southern hemisphere. Label each diagram with the term 'high' or 'low'.

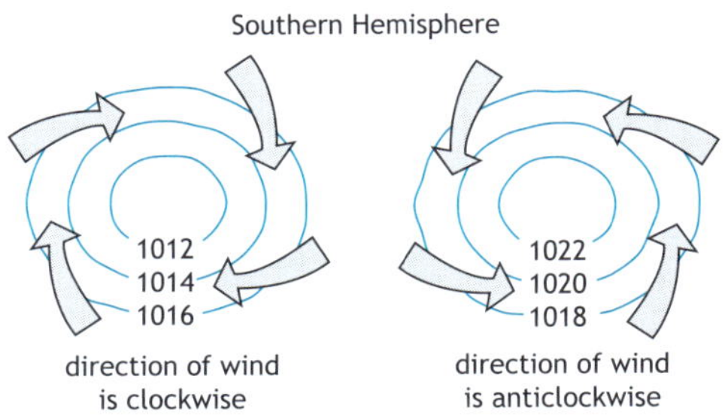

Question 7

The Bureau of Meteorology classifies cyclones according to their severity into categories from 1 (weakest) to 5 (strongest).

Study the table below.

Category	Average wind (km/h)	Strongest gust (km/h)	Central pressure (h/Pa)
1	63–90	less than 125	more than 985
2	90–120	125–170	985–970
3	120–160	170–225	970–945
4	160–200	225–280	945–920
5	more than 200	more than 280	below 920

a The following information relates to cyclone Tracy: gusts of 230km/h, minimum pressure of 940 hPa, average wind speed of 160 km/h. What category was cyclone Tracy (Darwin 1974)?

b The following information relates to cyclone Winifred: gusts of 170 km/h, minimum pressure of 960 hPa, average wind speed of 130 km/h.

What category was cyclone Winifred (North Queensland 1986)?

Question 8

The graph below shows how Australia's wheat yields have fluctuated with the Southern Oscillation Index. Explain the relationship between wheat yields and the SOI.

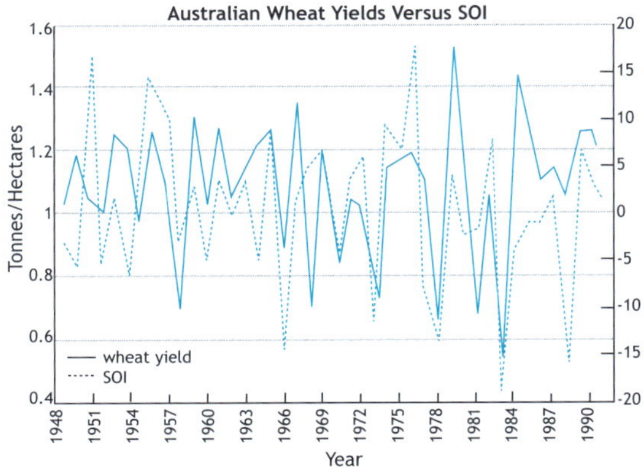

Question 9

Study the following graph of the progress of a typical El Nino event over 18 months.

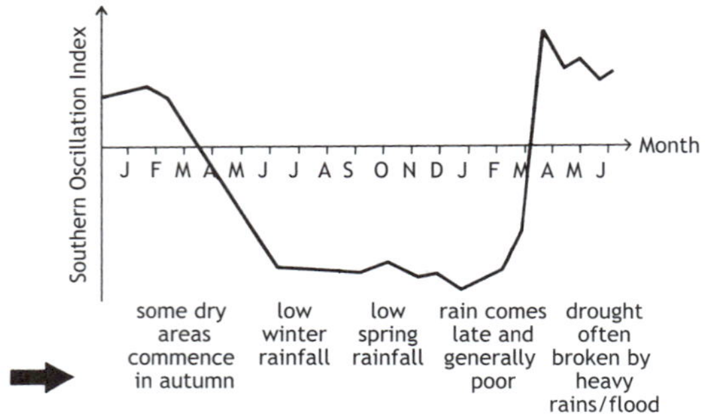

a In which month does a negative SOI begin? When does it end?

____

____

b How many months experience a positive El Nino event?

____

c What weather features occur in the negative SOI period?

____

____

____

Question 10

Complete the information below by adding the correct words.

When air begins to move it is called ________. Air ________ if it is ________ and falls if it is ________. Air also moves from areas of ______ pressure to areas of ______ pressure. Isobars which are close together indicate ____________ winds. Winds are deflected to the ______ in the northern hemisphere and to the _______ in the southern hemisphere. This is called the _________ ________.

Question 11

a Which of Newton's laws do the following situations below support?

____

____

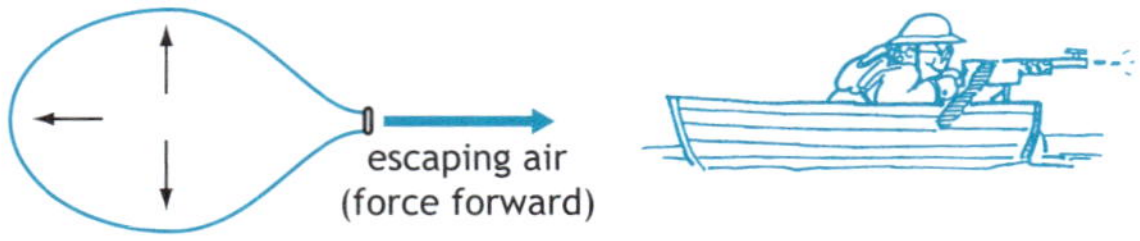

b In which direction will the balloon move?

____

c In which direction will the boat move?

____

Question 12

This bicycle is travelling at a constant speed but there are still forces acting on the bicycle.

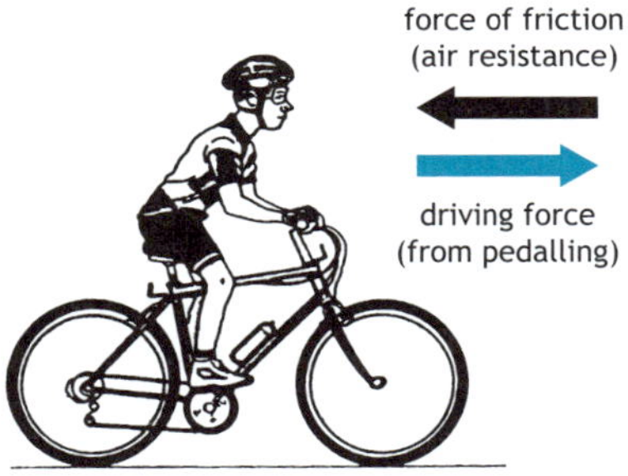

a What are the two forces affecting the speed of the bicycle?

____

____

b Why is the bicycle going at a constant speed?

____

c What forces do you need to change the way something moves?

____

Question 13

A projectile is an object fired by force. Students built the projectile launcher below.

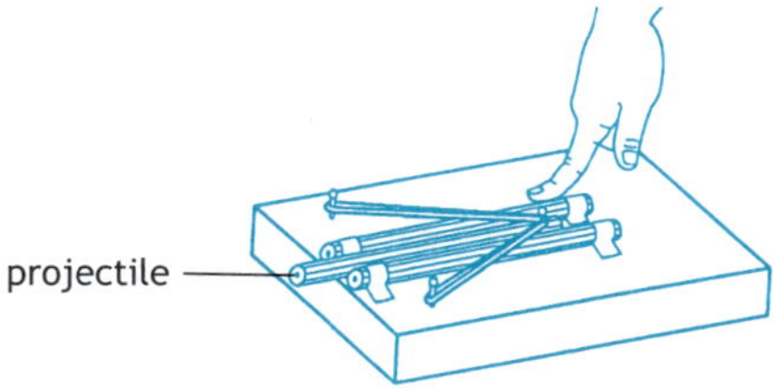

a How fast will the projectile move when fired, in km/hr, if it moves 3 metres in half a second? (There are 3600 seconds in one hour.)

____

b What will happen if the students use three elastic bands?

____

c What will happen if they use a heavier projectile?

d Which of Newton's Laws explains the action of this machine?

Question 14

Popcorn and rockets have a lot in common. When you heat the starch in the corn you also heat the water in it. The water turns to steam.

a What happens to the popcorn?

b Which of Newton's Laws explains this reaction?

c How is popcorn like a rocket?

Question 15

Motor car companies conduct crash tests on cars so they can make them safer for the passengers.

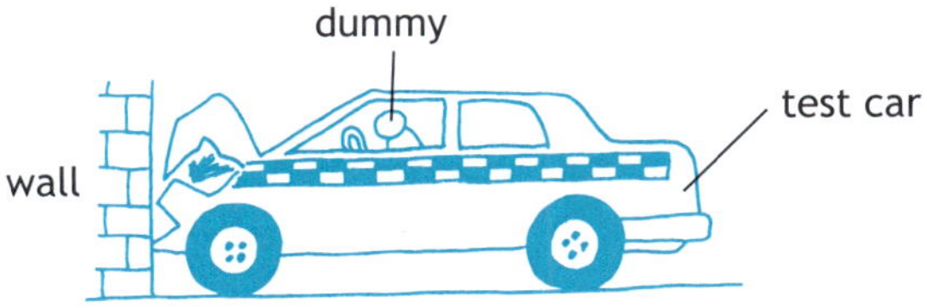

a Which of Newton's Laws are the companies working with?

b Complete the following explanation:

Moving objects need a ________ to stop them as their ________ tends to keep them moving in the ________ direction. The force from the ________ must be big enough to slow the passengers down when the car hits the wall.

Question 16

Students were experimenting with Newton's First Law. They were practicing putting backspin on a ball.

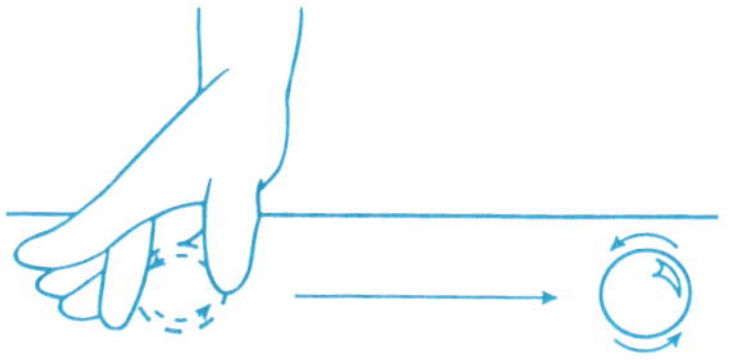

What force counteracts the motion forwards and makes the ball return to the sender?

Question 17

Using your knowledge of force, mass and acceleration together with Newton's First Law, predict what will happen below.

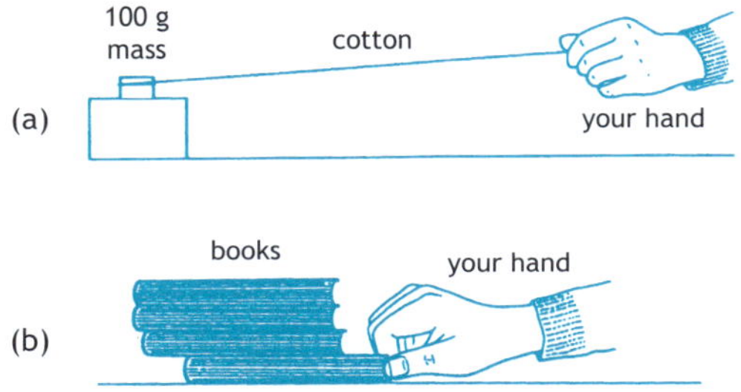

a What will happen if you pull the cotton slowly?

b What will happen if you pull the cotton hard suddenly?

c What will happen if you pull the book slowly?

d What will happen if you pull the bottom book out suddenly?

Question 18

An engineer needs to know how much a long steel beam will sag under a load.

The table below shows some results.

Load (N)	Sag (cm)
1000	2.0
2000	4.0
3200	6.6
4400	8.8
5200	10.4
6500	13.4

a Plot the graph of the sag against the load.

b One of the sag measurements is wrong. Which is it and what should it be?

____

c What would be the sag for a load of 4500 N?

____

d What load would give a sag of 52 mm?

____

Question 19

A child is riding on a roundabout in a playground.

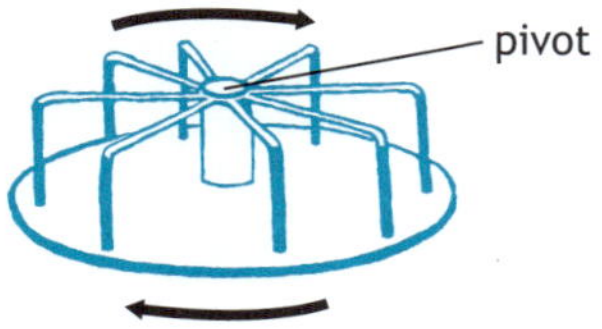

a State Newton's First Law.

____

____

____

b Is the roundabout moving in a clockwise or anticlockwise direction?

____

c What force is keeping the child on the roundabout?

____

d If the child steps off this moving roundabout which way will he have to run to avoid falling over?

____

____

e What force will slow the roundabout down?

____

Question 20

Three radioactive substances have to be stored safely. Details of the substances are given in the table below.

Substance	Half-life	Type of radiation
A	4000 years	Alpha
B	5 years	Beta
C	166 years	Gamma

a Which of the following containers would you use for each substance?

i aluminium ______

ii paper ______

iii thick lead ______

b Complete the table below for substance B above.

Date	Mass of radioactive substance left
1 January 1990	8 kg
1 January 1995	
1 January 2005	

Question 21

A Geiger counter was used to measure the activity (counts per minute) from a radioactive sample in a laboratory. Over this period, the background radiation was regularly measured at 4 counts/minute. The table of results follows.

Time in years	Activity in counts/minute	Activity due to the sample alone
0	124	120
1	80	
2	52	
3	34	
4	23	
5	16	
6	12	8

a Complete the table giving the activity of the sample alone.

b Explain what is meant by background radiation.

c Draw a graph, plotting the values for the sample alone against time, and join them with a smooth line.

d Find the half-life of the substance from your graph.

Question 22

Your body is receiving background radiation all the time due to natural sources. The amount of radioactivity you receive depends on things like where you live and what your job is. The following graph shows the background radiation in Britain.

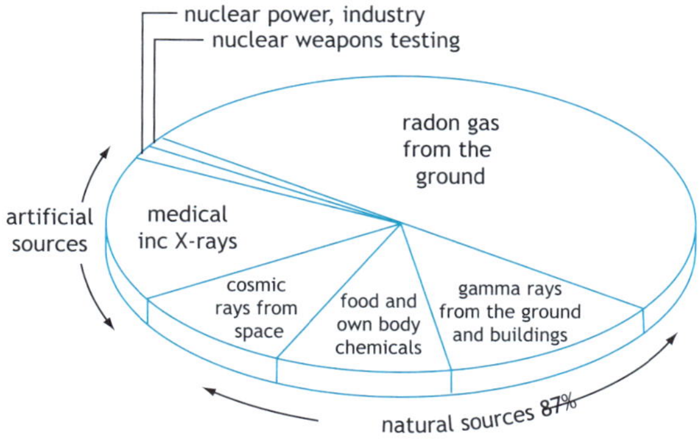

a What is the biggest single source of background radiation in Britain?

b Which industry generates a significant amount of radiation?

Question 23

All living things take in some radioactive carbon in photosynthesis. When they die they stop taking in carbon and the radioactive carbon decays with a half-life of 5700 years.

A fresh bone gives a radioactive count of 100 counts/minute. A bone of the same mass from a skeleton gives 30 counts/minute. The background count rate is 20 counts/minute. How old is the bone?

Question 24

Arrange the following terms in the correct order to describe the way a coal-fired power station works.

steam, boiler and furnace, fossil fuel, generator, turbine, electricity

Question 25

The following pie graph shows Australia's non-renewable fuel resources.

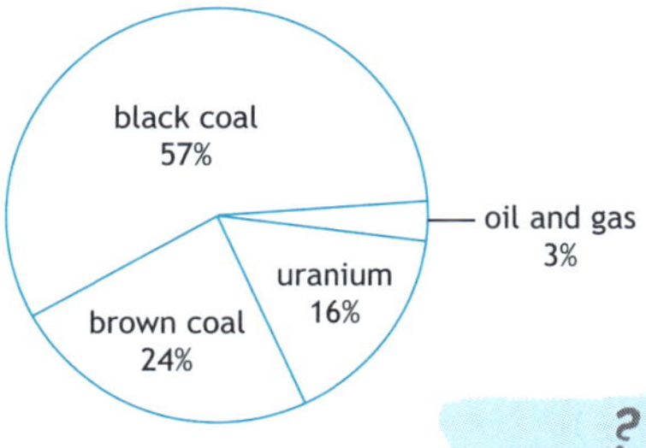

a What is Australia's smallest energy reserve?

b What is Australia's largest energy resource?

c Do we use uranium in Australia?

Question 26

The following map shows an industrial area. The prevailing wind is from the west. There is a mountainous region nearby from which a river flows. The major road and rail links are shown.

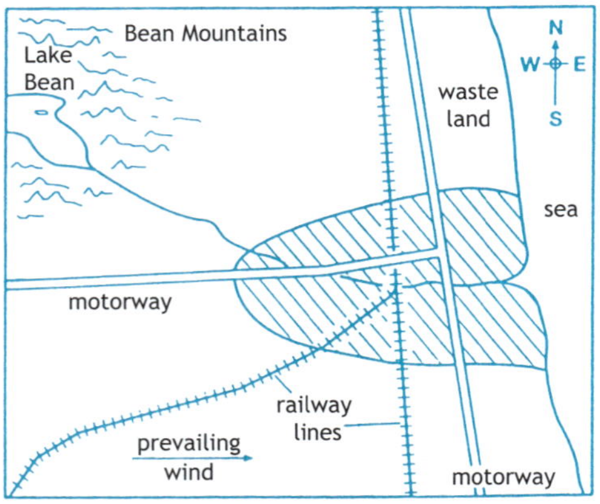

A power station is to be built to supply electrical energy to the region for domestic and industrial use. The choice is between a coal-fired power station, wind turbines and a hydro-electric scheme.

Complete the table with one disadvantage and one advantage of each of these options.

Power option	Disadvantage	Advantage

Question 27

Complete the following sentences.

We sometimes find the energy sources we need by burning ________ ________. The energy stored in fuels is called __________ energy. When fuels are burned energy is transferred to the surroundings as _______ energy.

Question 28

Coal, gas and oil were formed millions of years ago.

a What is coal made from?

b What other sorts of fuels were made in the same way?

c Why are coal, gas and oil called fossil fuels?

Question 29

Add the correct words.

Some energy sources will last forever because the energy is constantly being __________. These energy sources are described as ______________.

However, some sources will eventually run out. These are described as _______________ energies.

Question 30

Study the pie graph below, which shows the amount of carbon stored in parts of the Earth.

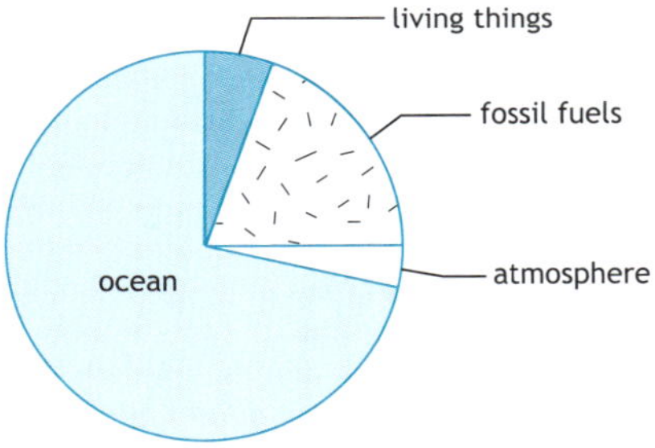

According to this information which statement is incorrect?

a More coal is found in fossil fuels than in the atmosphere.

b Living things contain the most carbon.

c About one quarter of the carbon is found in fossil fuels.

d About 70% of the carbon is found in the ocean.

Question 31

Study the following drawings and label each as renewable or non-renewable.

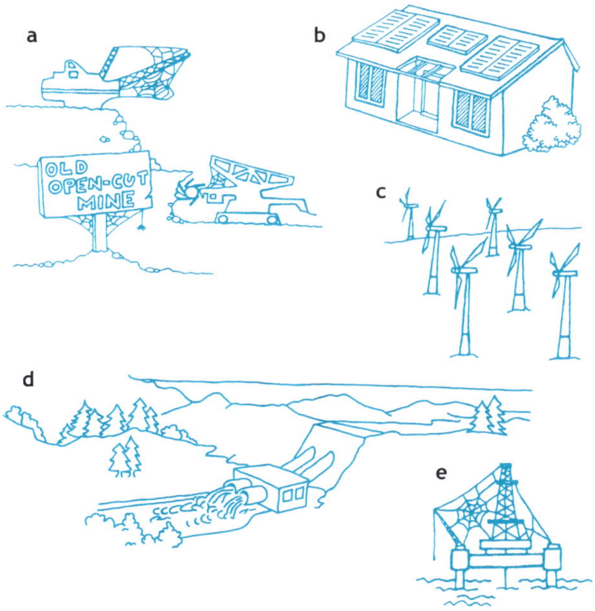

Question 32

Look carefully at the diagrams below of the sun's path over an energy-efficient house in both summer and winter. The house has extended eaves and a deciduous tree outside the window.

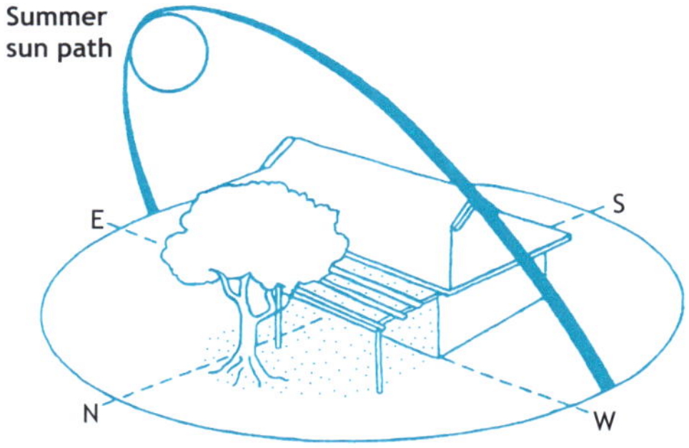

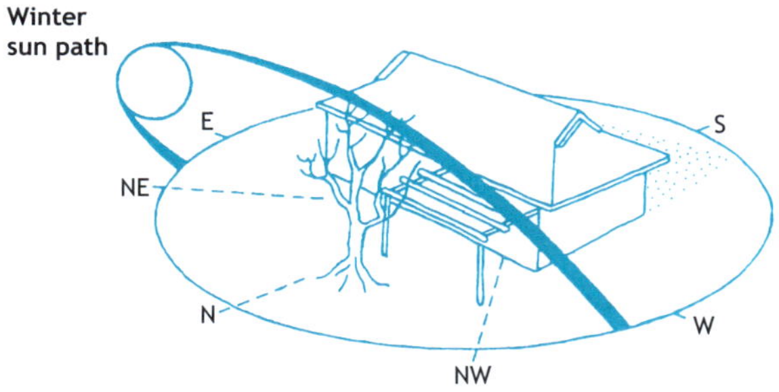

a What role does the tree play in conserving energy?

____

____

____

____

b What role do the eaves play in conserving energy?

____

____

____

____

Answers

Chapter 1

Science skills page 5

1 a Triassic
  b Late Carboniferous and Permian
  c bony fish, birds and mammals
  d cartilaginous and bony fish
  e jawless fish
  f 180 million
  g Mesozoic era
  h dinosaurs (reptiles)
2 A close structural resemblance suggests that they may have a common ancestor.
3 a Devonian period b Graptolite and Trilobite
4 e, g, b, d, h, f, a, c
5 a (4) – in the early Cretaceous period.
  b (5) – in the Tertiary period.
6 a 1 412 900 b insects and flowering plants
  c land d 30 800
7 a 10 000 years ago b about 25%
  c almost 100% d Africa
  e Madagascar and New Zealand
8 a d b c
9 a = 2 b = 4 c = 3 d = 1
10 a graptolites, trilobites and goniatites
  b The rock was laid down between 510 and 225 million years ago.
  c Rocks laid down about 500 million years ago.
  d 65 million years ago, after the dinosaurs.
  e conifers and primitive ferns
11 B
12 Australopithecines
  Lucy
  Homo
  Neanderthal man
  Homo sapiens
13 a They were dying in large numbers.
  b The myxoma virus was released.
  c natural selection
  d The population had adapted and were becoming immune to the virus.
14 a c, a, b, d
  b One toe only (the hoof).
  c natural selection

Revision questions page 9

1 caste
2 a 1. A large number of organisms that were alive in the past have either not been preserved or have not been discovered yet.
  2. It is rare to find a complete fossil and scientists must infer the structure of the whole organism from the parts found.
  b 1. Some fossils can be used to tell the age of the rock in which it was fossilised, as the organism lived only at a particular time in history.
  2. Fossils can show the rise and fall of populations throughout the history of the Earth.
3 index, Palaeozoic era, extinct
4 a Mesozoic b Palaeozoic c Permian
  d Devonian e Precambrian
5 reptiles, 200, extinct, 65, asteroid, climate, adapt
6 a Hominids
  b They walk on two legs and have large brains.
  c It freed their hands and gave them the ability to hold tools.
  d 15 million years
7 Charles Darwin, Beagle, plants, animals, evolution, similar, Origin, Species
8 a The species evolves.
  b extinction
  c The numbers of a species fall to such a low level that the organism may become extinct.
  d habitat destruction
9 a (A) = mammal (B) = reptile
  b (A) = hunting by humans (B) asteroid
  c Amphioxis
10 mould
11 a Evolution is the gradual change in the characteristics of a population over many generations. This takes a long period of time and results from changes in the environment.
  b An adaptation is a special feature which gives an organism a better chance of survival.
  c The 'fittest' individuals reproduce and pass on any special features they may have to their offspring.
12 a mineralisation b petrifaction
13 a ice ages b evolution
14 a 500 million years
  b Cambrian era
  c A relatively sudden increase in atmospheric oxygen levels.
15 a Eras, periods and epochs. b era
  c era = Cenozoic, period = Jurassic
16 a Because when Silurian rocks were laid down most of the organisms on the planet were soft-bodied.
  b Because this land must have been covered by the sea in the past.
  c fish, amphibians, dinosaurs, mammals
17 c
18 radioactive dating, age, decay, carbon-14, fossils
19 a Immigration of individuals with new genes into a population.
  Recombination of genes during meiosis.
  Mutation.
  b Because it enables living things to change slowly through the process of evolution.
20 1. No life.
  2. Many sea creatures.
  3. Amphibians, ferns and conifers flourish.
  4. First reptiles appear, many large insects.
  5. Reptiles thrive, first mammals.
  6. Giant dinosaurs, first birds, first flowering plants.
  7. Dinosaurs die out.
  8. Mammals, modern flowering plants.
  9. Early hominids.
  10. Humans dominate, domestic animals.

Chapter 2

Science skills page 18

1 a

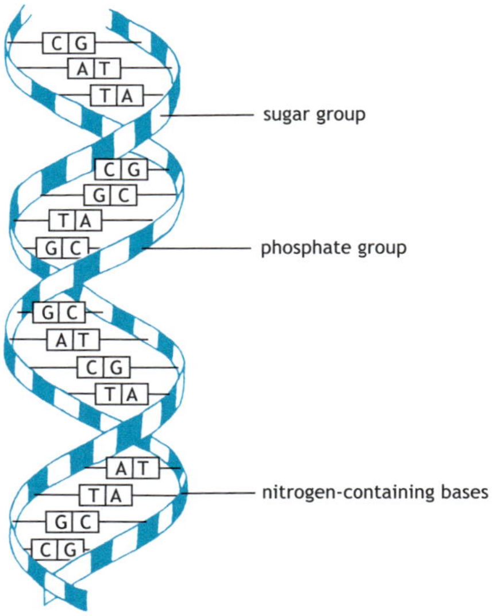

  b C – Cytosine
  G – Guanine
  A – Adenosine
  T – Thymine

2 a replication
b DNA unzips itself to form two separate strands. The two strands form a complementary copy of themselves using free bases in the cell. Two new DNA molecules are produced.
c G-C-C-T-A-A

3

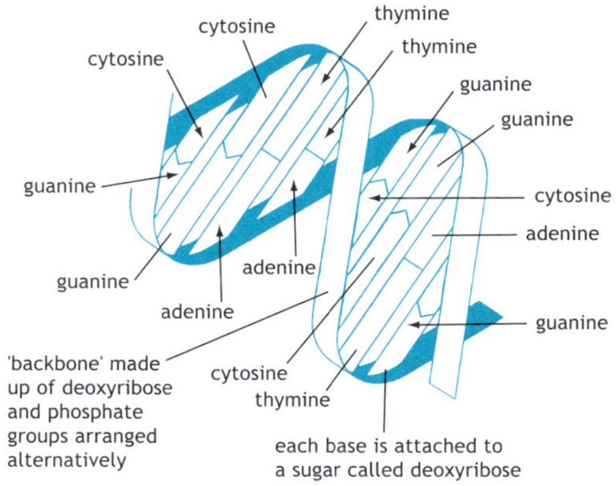

4 a genes b transcription c DNA
d DNA e translation
f ribosomes, messenger, transfer, codons, amino acids, protein
5 a one red and one white b Rr
c R and r d 3:1
6 a SS, ss b Ss c All smooth skinned.
d 3 smooth skinned to 1 wrinkled skinned. e SS, Ss, Ss
7 a Male = LL, Female = ll
b All the offspring will have the genotype Ll and they will all have long floppy ears.
8 a Possible genotypes: LL, 2Ll, ll
b Possible phenotypes: long floppy ears or short ears.
c Ratio of genotypes is 1:2:1.
9 a i Male = NN Female = nn
ii Male = Nn Female = nn
iii Male = Nn Female Nn
b i genotype: all offspring will have a genotype of Nn.
phenotype: all offspring will have a phenotype of normal fur.
ii genotype: there will be two – Nn and nn. The ratio will be 1:1.
phenotype: there are two – normal fur and ridgeback. The ratio is 1:1.
iii genotype: there are three – NN, Nn, nn. The ratio is 1:2:1.
phenotypes: there are two – normal fur and ridgeback. The ratio will be 3:1.
c The second mating, a heterozygous male crossed with a ridgeback female. In this mating the chance of getting a ridgeback offspring is 50%, whereas with the third mating the chance it is only 25% and with the first mating the chance is zero.
10 Red Bull White Cow
RR rr
First generation = all red with a genotype of Rr.
Rr Rr
Second generation = ratio of 3 red : 1 white
RR Rr rR rr
11 c
12 a brown
b Yes, but only if the partner has two recessive genes.
13 Alice, Irene and Alexandra.
14 The wolf has bases similar to the wild dog but unlike the domestic dog. (Note sections 190, 200 and 210 on the chromosome.)
15 a, b, c, d

Revision questions page 22

1 Codon A three-base group used in the genetic code.
DNA Deoxyribonucleic acid.
Chromosome A thread-like structure that contains all or part of a cell's DNA.
Gene A basic unit of heredity that carries the instructions needed to make a particular protein.
2 replication
3 a double helix b bases
c protein synthesis d messenger RNA and transfer RNA
4 a James Watson and Francis Crick
b Gregor Mendel
5 DNA, RNA, DNA, RNA, DNA
6 a phosphate groups
ribose sugar
nucleic acids
b Chromosomes are strands of DNA and genes are small pieces of the DNA molecule. A chromosome is made up of lots of genes joined together.
c A gene is a segment of a DNA molecule with a particular sequence of three nitrogenous bases. These triplets are coded in such a way that they each represent a specific nucleic acid.
7 a The chemical code used by DNA and RNA.
b Adenine, guanine, thymine and cytosine.
c It tells the cell how to convert the DNA's sequence of bases into amino acids.
d amino acid
8 a phenotype b heredity
c allele d genotype
9 alleles
10 a evolution b mutation
11 a BB b Bb c black
d homozygous e heterozygous
12 a Tt x Tt b (T or t) x (T or t)
c TT, Tt, tT, tt d tall, tall, tall, short
13 Homozygous cell A cell with identical alleles.
Heterozygous cell A cell with one dominant and one recessive allele.
Linked alleles Alleles that are close to each other on the chromosome and are inherited together.
Sex-linked alleles Alleles carried by the sex chromosomes.
Mutation A change in a cell's genetic material.
14 locus
15 Human Genome, adenine, guanine, cytosine, thymine, phosphates, sugar

Chapter 3

Science skills page 30

1 a

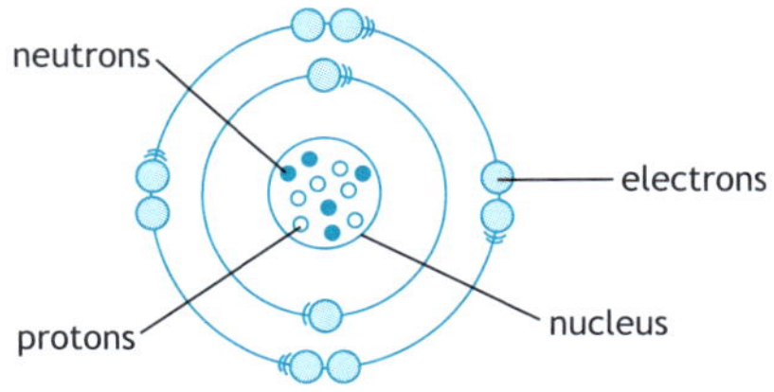

b Solar System Model
c Niels Bohr
2 a atomic number = number of protons
= number of electrons
b mass number = number of protons + number of neutrons
c Number of neutrons = mass number – atomic number
3

Element	Symbol
Hydrogen	H
Carbon	C
Sodiuum	Na
Zinc	Zn
Chlorine	Cl
Oxygen	O

4 a It becomes negatively charged.
b It becomes positively charged.

c

Case	Protons	Electrons	Overall charge
neutral atom	1	1	0
gain an electron	1	2	–ve
lose an electron	1	0	+ve

5 a

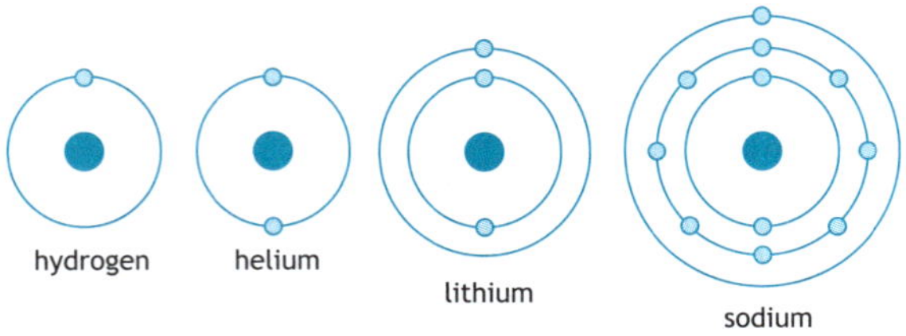

b H, He, Li, Na
c hydrogen, lithium and sodium
d Because they have only one electron in the outer shell.

6 a

Substance	How many types of atoms	Element or molecular compound
oxygen	1	element
water	2	molecular compound
glucose	3	molecular compound
hydrogen	1	element

b

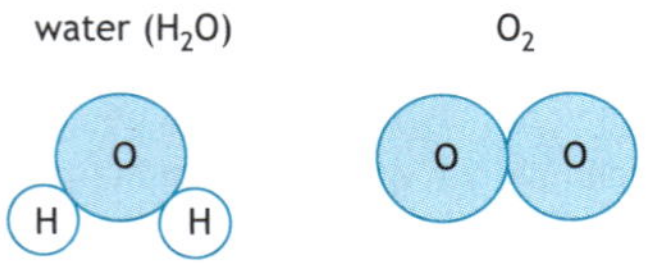

7 a A and D
b B and C – more than one type of atom must be present.

8

Symbol	Element
H	Hydrogen
He	Helium
Li	Lithium
C	Carbon
N	Nitrogen
Na	Sodium
Zn	Zinc
Ar	Argon
Fe	Iron

9 a For answer see base of page
b Column 18
c Column 1
d Column 17

e Column 17, because the elements in column 16 must gain 2 electrons to fill the outer shell, whereas those in column 17 need only one.
f The ones to the right of the zigzag line.
g Sc, Ti, V, Cr, Mn, Fe, Co, Ni, Cu, Zn
h Column 17

10 a

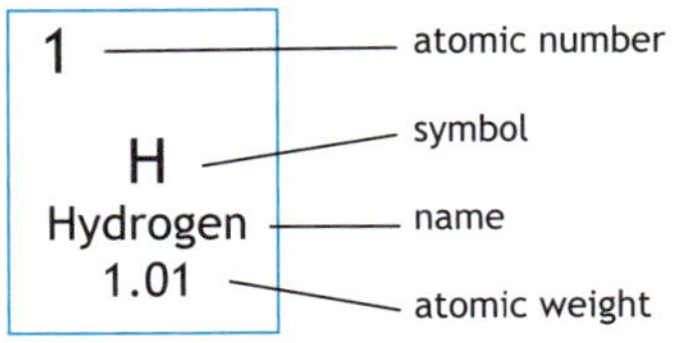

b Atomic number = 20
Atomic mass = 40.08

11

Element	Likely electron movement	Does it become a cation or anion?
sulpher	gain 2	anion
magnesium	lost 2	cation
calcium	lost 2	cation
argon	no ion formed	inert
bromine	gain 1	anion

12 a

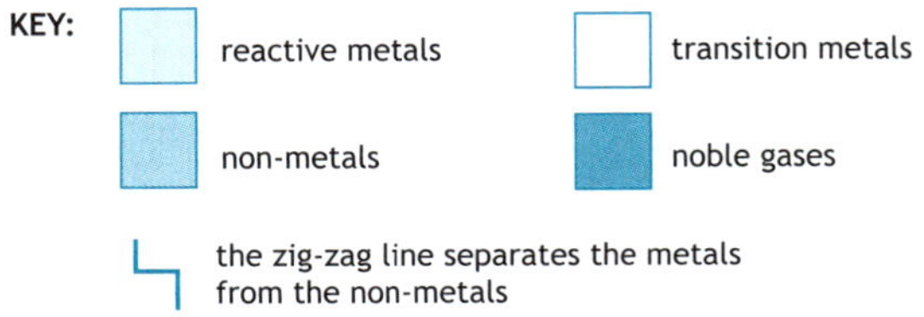

b 11, 8, 29
c chlorine, carbon, sodium
d 2, 8, 5

13 a helium, neon, argon, krypton
b They belong to the group called the noble gases.
c They are inert, they outside shell of each of the types of atom is full.

14 b

15 a yes b gas c it glows

Revision questions page 34

1

Subatomic particle	Electric charge
electron	negative
neutron	neutral
proton	positive

Question 9 a

1	2	3	4	5	6	7	8	9	10	11	12	13	14	15	16	17	18
1 H 1																	2 He 4
3 Li 7	4 Be 9											5 B 11	6 C 12	7 N 14	8 O 16	9 F 19	10 Ne 20
11 Na 23	12 Mg 24											13 Al 27	14 Si 28	15 P 31	16 S 32	17 Cl 35.5	18 Ar 40
19 K 39	20 Ca 40	21 Sc	22 Ti	23 V	24 Cr	25 Mn	26 Fe 56	27 Co	28 Ni	29 Cu 63.5	30 Zn 65	31 Ga	32 Ge	33 As	34 Se	35 Br 80	36 Kr

2

Factors differentiating atoms	Factors differentiating substances
1. The type of atom.	1. Type of element present.
2. Number of each type of particle.	2. How many atoms of each are present.
3. Patterns in which the particles are combined.	3. Patterns in which the atoms of each element are combined.

3 a false b true c false d true
e true f true g false h true
i false j true
4 a atom b compound
c molecule d element
5 It loses two electrons.
6 a The Periodic Table
b It consists of:
- seven periods or rows of increasing atomic weight
- eighteen columns.

c It is a summary of the atomic weights, numbers and symbols of the elements and it groups elements of similar properties into columns for easy identification.
7 a 11 b 11 protons, 11 electrons, 12 neutrons
c one
8 a cation b ion c anion
9 a There are more metals.
b

Group of elements	Properties
metals	High lustre, conduct electricity, malleable
non-metals	lack lustre, do not conduct electricity
noble gases	unreactive and glow brightly when electricity is passed through them

10 properties, reactive, alkali, halogens
11
- The atom has six protons.
- The atom has six positive particles.
- The atom must have six negative particles.
- The number of neutrons must be 6 (the mass number = 12 – 6).

12 They would be the same.
13

1	2	3	16	17	18
H^+			O^{2-}	F^-	nil
Li^+	Be^{2+}		S^{2-}	Cl^-	nil
Na^+	Mg^{2+}	Al^{3+}	Se^{2-}	Br^-	nil

14 Compound Pure substance made of more than one element.
Element Pure substance made of only one type of atom.
Atom Smallest bit of an atom.
Molecule A group of atoms.
15 The properties of a substance depend on how many atoms are present and how they are arranged.

Chapter 4

Science skills page 42

1 a Eurasian plate
b Himalayan mountains
c earthquakes
d The Nazcan plate is pushing the South American plate. This is causing mountain building and earthquakes.
e Indo-Australian plate
f No, because there are no plate boundaries crossing Australia.
g The Pacific plate and the Nazcan plate.
2 a Pre-Cambrian era
b It was weathered.
c On the edges of the continent's core.
d Palaeozoic era
e mountain folding
3
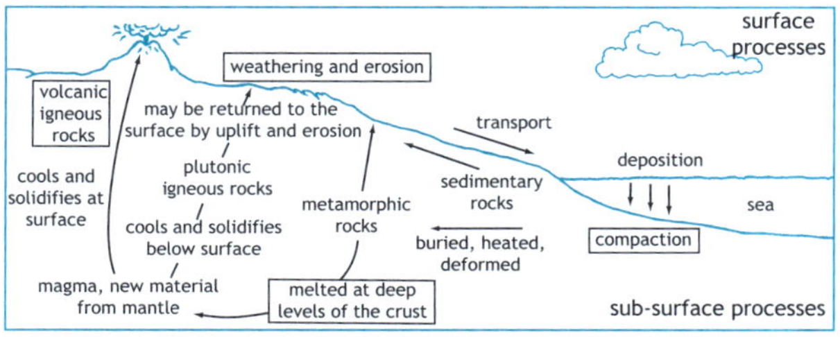

4 volcanoes and earthquakes
5 i = d
ii = b
iii = a
iv = c
6 Volcano A vent in the Earth's crust from which lava and other materials are ejected.
Magma Hot molten rock generated within the Earth.
Granite A common, coarse grained, light coloured intrusive rock.
Lava Hot molten rock that reaches the surface of the Earth.
Sill A horizontal body of intrusive rock forced between layers of enclosing rock.
Laccolith An igneous intrusion that has arched up the rock strata into which it was forced, forming a lensshaped body with a flat floor.
7
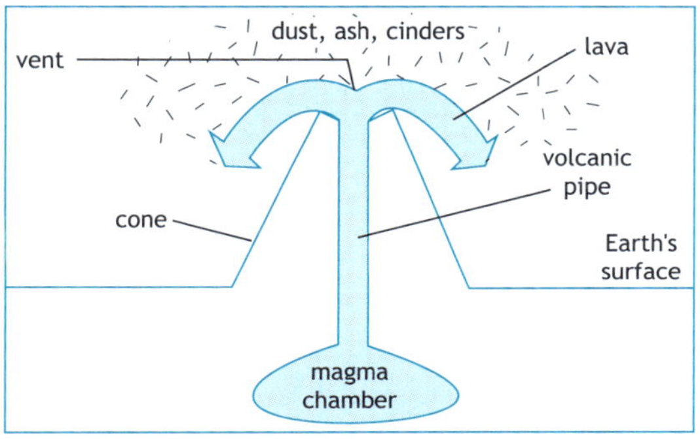

8 a = i b = iii c = ii
9
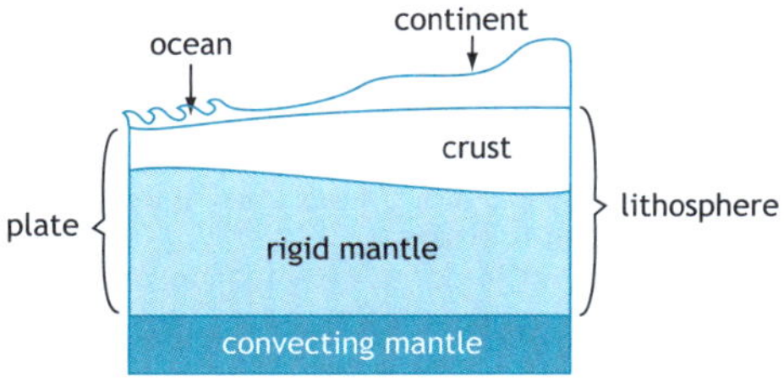

10
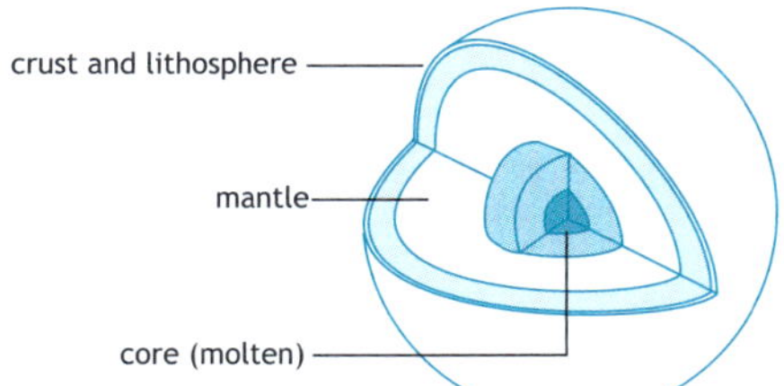

11 a rhea
b kiwi
c Some groups of organisms have a southern hemisphere distribution. DNA comparisons indicate that these flight less birds are very closely related, suggesting that the landmasses on which they are found were once connected.
d Geology and fossil records.

12

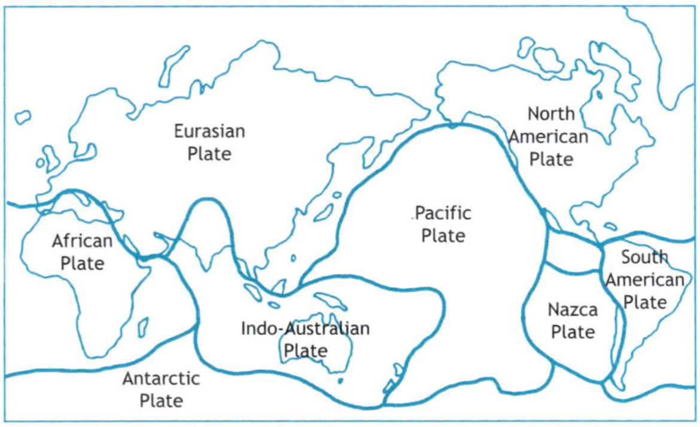

13

Landform	Boundary
The Great Dividing Range	C
The Himalayan Mountains	D
Andes in South America	D
Hawaiian Islands	C
Mid-Atlantic Trench	C

14

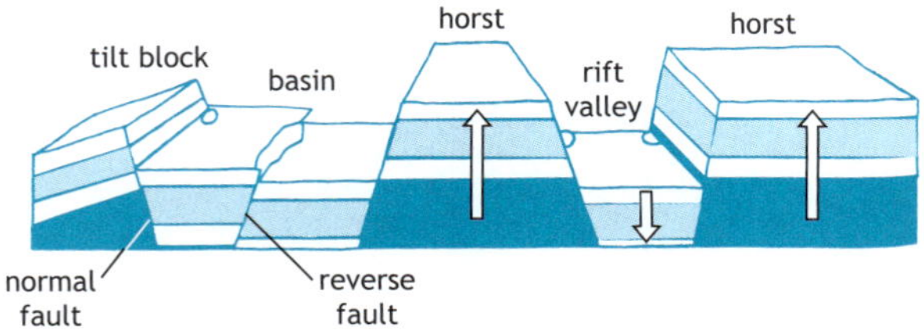

15 a 80° and 60° south
b 90 million years
c about 78° south
d about 44° south
e The climate has gradually got warmer.

Revision questions page 46

1 a A – Constructive plate boundary
B – Destructive plate boundary
b A – Where two plates move apart, new material wells up as a lava flow.
B – Where two plates collide head-on, one slides under the other returning material to the mantle.
c subduction
2 a Rising magma is stored under pressure in the magma chamber and upwells along fault lines and other zones of weakness to the surface.
b Where two plate boundaries are pushing on one another.
c Ring of Fire
d basalt
e A vertical igneous intrusion across strata.
3 tsunami, earthquakes, shallow, higher
4 a The theory of continental drift.
b Antarctica must still have been connected to Australia when marsupials evolved, but that they died out. Perhaps because Antarctica's climate got too cold as it drifted south after it split from Australia.
5 a Pangaea
b Laurasia and Gondwana
c A, B, D, C
6 a Sometimes a narrow section of magma burns through the crust. This section remains stationary while the plate moves over it.
b A hot spot.
c The Great Dividing Range.
7 A. batholith B. laccolith C. sill D. dykes
8 plates, tectonic, continents, earthquakes, mountain, mid-Atlantic trench, destroyed
9 a C b A
c fault line d folded
10 a igneous
b Molten magma cooling below the surface.
c They have been folded.
d Magma was forced up through a crack in the Earth's crust.
e Erosion of the overlying layers of rock.
11 Antartica, Asia, India, Africa, South America
12 a horst or tilt block b rift valley
13 a India
b folding
c San Andreas Fault
d The Great Rift Valley
e volcanic plug, crater lake, cinder cone, caldera, lava flow, etc
14 a the eastern States b the Great Artesian basin
15 Igneous, obsidian, pumice, lava, quickly, surface, slowly, granite

Test A page 49

1 a Because they are of the same material and have the same fossils.
b 5 and 12, as they also contain the same fossils.
c 10
d 4
2 c
3 a mass extinctions b 5
c Between the Permian and Triassic periods.
4 1. That all living things vary.
2. That they can pass on their characteristics.
3. That they are involved in a struggle for survival.
5 a & b

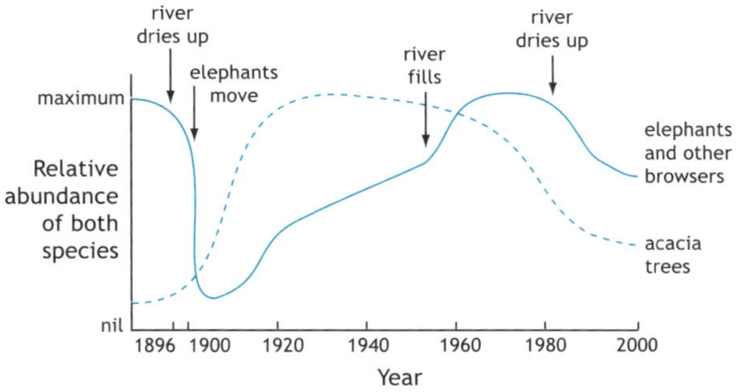

c Because there were no elephants to browse on the acacias.
d The river filled up and the elephants came back.
6

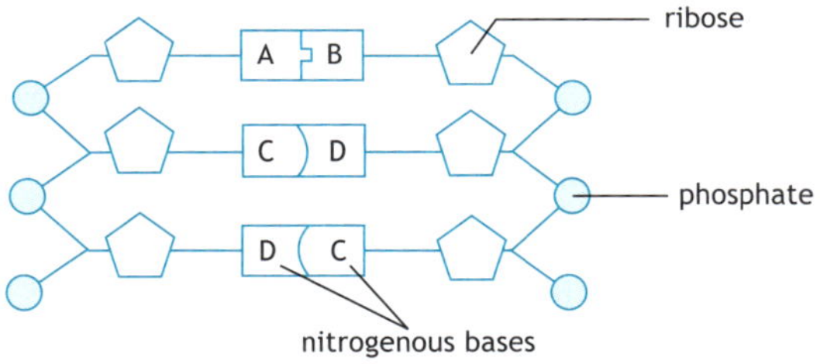

7

Nucleotide	Full name	Where in the cell	Function
DNA	Deoxyribo-nucleic acid	nucleus of a cell	contains the genetic code
RNA	Ribonucleic acid	nucleus and cytoplasm of a cell	carries the genetic code

8 e, b, d, a, c
9 a BB b Bb c all black d 3 black : 1 white
10 a 50% b 25%
11 a 3 long horned : 1 short horned b HH, Hh, hh
12 a homozygous, homozygous
b homozygous, heterozygous

13 b

14

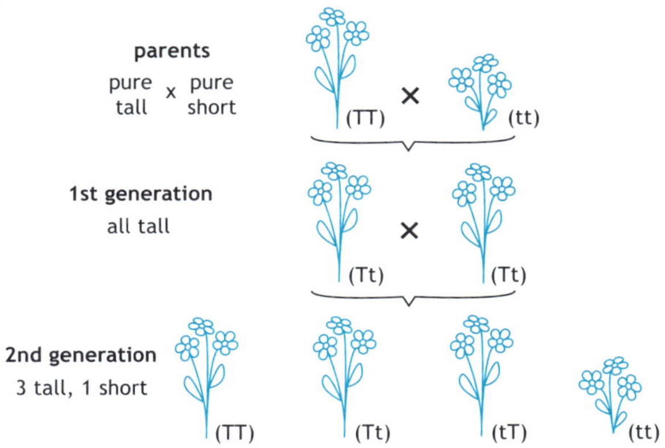

15 No, as all individuals must carry a dominant gene.

16

Atomic number	Mass number	No. of protons	No. of neutrons	No. of electrons
14	28	14	14	14
11	23	11	12	11
19	39	19	20	19
7	14	7	7	7

17

	1	2	3	4	5	6	7	8	9	10	11	12	13	14	15	16	17	18
1	1 H																	2 He
2	3 Li	4 Be											5 B	6 C	7 N	8 O	9 F	10 Ne
3	11 Na	12 Mg											13 Al	14 Si	15 P	16 S	17 Cl	18 Ar
4	19 K	20 Ca	21 Sc	22 Ti	23 V	24 Cr	25 Mn	26 Fe	27 Co	28 Ni	29 Cu	30 Zn	31 Ga	32 Ge	33 As	34 Se	35 Br	36 Kr
5	37 Rb	38 Sr	39 Y	40 Zr	41 Nb	42 Mo	43 Tc	44 Ru	45 Rh	46 Pd	47 Ag	48 Cd	49 In	50 Sn	51 Sb	52 Te	53 I	54 Xe
6	55 Cs	56 Ba	57 La	72 Hf	73 Ta	74 W	75 Rc	76 Os	77 Ir	78 Pt	79 Au	80 Hg	81 Tl	82 Pb	83 Bi	84 Po	85 At	86 Rn
7	87 Fr	88 Ra	89 Ac	104 Rf	105 Db	106 Sg	107 Bh	108 Hs	109 Mt	110 Ds	111 Rg	112 Cn		114 Fl		116 Lv		

18

Element	Symbol	Atomic number	No. of protons	Atomic mass
lithium	Li	3	3	6.94
carbon	C	6	6	12.01
chlorine	Cl	17	17	35.45
barium	Ba	56	56	137.36
gold	Au	79	79	196.97

19 a Pt b Hafnium c 79 d 126.9

20 There are 45 atoms combined in a molecule of sucrose: 12 carbon atoms, 22 hydrogen atoms and 11 oxygen atoms.

21 acids, hydrogen, salts, magnesium chloride

22 Metals: potassium, sodium, magnesium, calcium
Non-metals: sulphur, bromine, iodine, chlorine
Examples: sodium chloride, potassium iodide

23 A the hydrogen ion
B the hydroxide ion
C the ammonium ion
D the nitrate ion

24 c

25 a sodium b negative c neutral

26

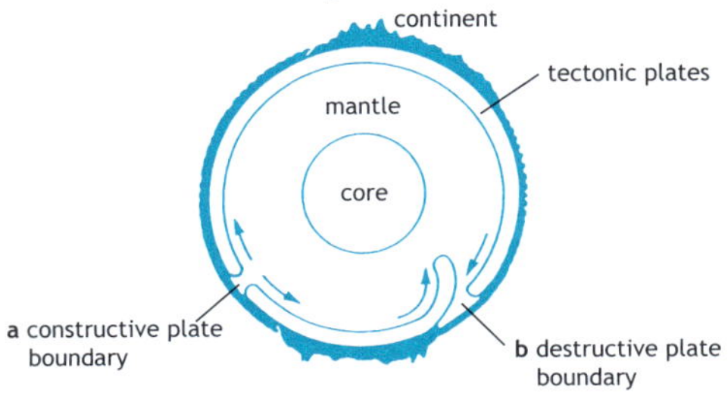

27 crust, plates, slowly, mantle, apart, mid-ocean ridges, subduction, slides, under, mantle, crust

28

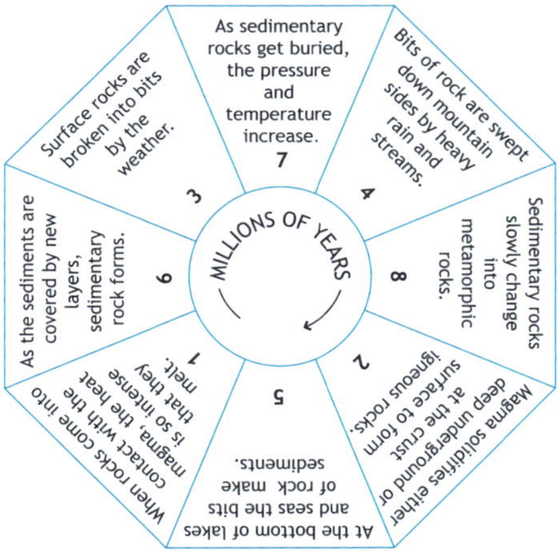

29 Earthquake — A sudden movement of plates in the Earth's crust, caused by the abrupt release of strain that has built up over time.
Basalt — A common fine-grained extrusive rock.
Crust — Earth's solid outermost layer.
Hot spot — A location where a stationary plume of magma rising from the mantle melts a hole in the Earth's crust, producing a volcano.
Seismograph — An instrument that measures the amount of ground movement produced by an earthquake.

30

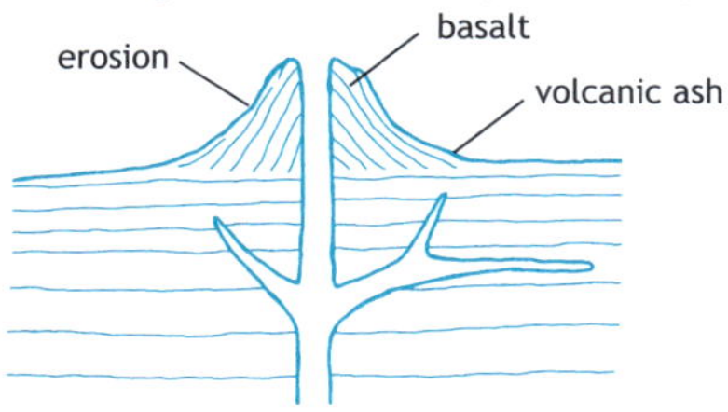

31 a Subsidence of the leading edge of a lithospheric plate into the mantle.

b

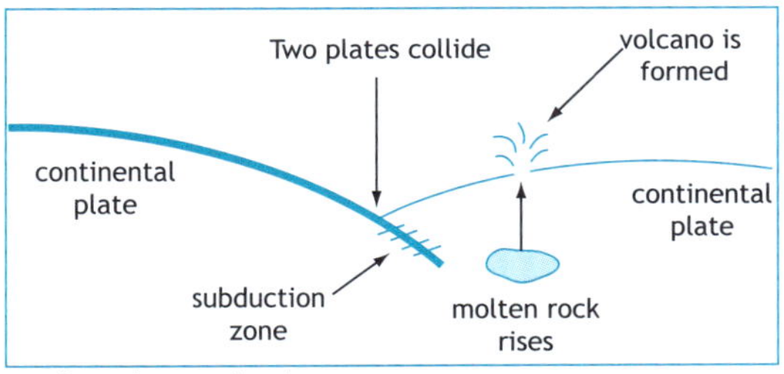

c earthquakes and volcanoes

32 a Richter scale b E c B d A and B

33 Composite cone — A cone formed from alternate layers of ash and lava.
Caldera — Steep-sided crater caused by a violent volcanic eruption.
Lava flow — Within rock layers under the ground surface it forms intrusive igneous rock.
Volcanic plug — The resistant core from an eroded volcanic

Chapter 5

Science skills page 59

1 a

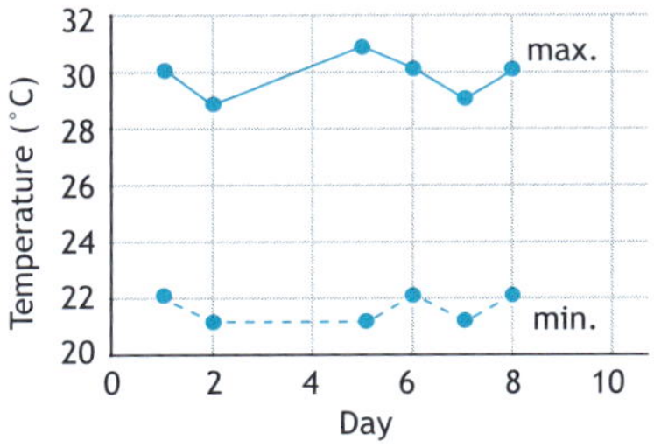

- b Day 5
- c Day 5
- d Day 9

2 a easterly
  b wind vane

3 a 8 mm
  b Thursday
  c 4 + 10 + 9 + 16 + 8 = 47 mm
  Average = 9.4 mm

4 a

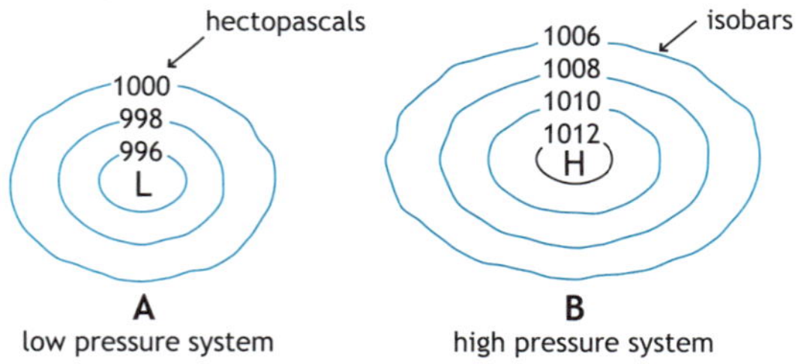

  b high pressure system (B)
  c atmospheric pressure

5 a cloudy and rainy
  b a gentle breeze – the isobars are not close
  c from left to right
  d fine with a hardly a breeze
  e from the north east

6 a & b

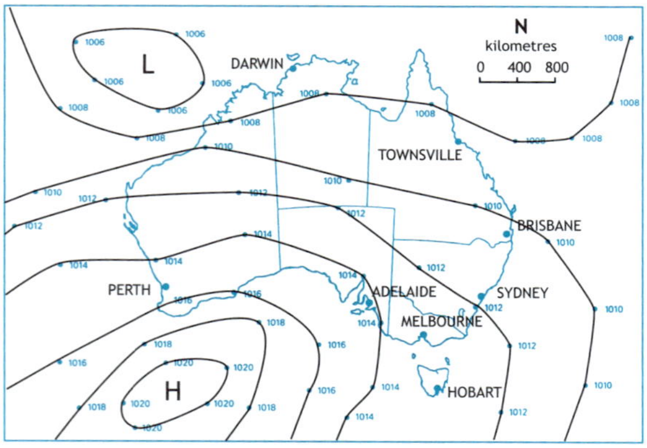

  c cloudy with rain approaching from the west

7 a cyclone
  b rain
  c clockwise

8 a A very wet year.
  b A drought year.

9 (B)

10 a i 52 km/h
  ii 24 km/h
  iii 9 km/h
  b i Moderate gale – whole trees in motion; inconvenience felt when walking.
  ii Moderate breeze – dust and loose paper move; small branches move.
  iii Light breeze – wind felt on face; wind vanes move.

11 a 22 °C
  b 71%
  c 67%

12 a cyclone
  b 25 km/h
  c 170 km/h, 0 km/h
  d 950 hectopascals

13 a Darwin
  b Darwin
  c Alice Springs
  d Perth

14 a Thunderstorm activity decreases as distance from the equator increases.
  b December
  c Winter

Revision questions page 64

1 water vapour, high, cirrus, stratus, cumulus

2

Climate	Average weather conditions of a particular place.
Weather	Atmospheric conditions in a particular place at a particular time.
Meteorology	The study of the atmosphere.

3 a air pressure
  b rainfall
  c wind speed
  d temperature
  e wind direction

4 a degrees Celcius
  b The areas are shaded.
  c It estimates wind speed by observing the effect the wind has on its surroundings.
  d The type of weather carried by the wind – eg a hot wind comes from blowing over hot land.
  e Strong winds and storms are on the way.

5

Thunderstorm	A storm accompanied by heavy rain, gusty winds and lightning.
Cyclone	A region of low atmospheric pressure accompanied by violent storm conditions.
Hurricane	A massive, spiralling low pressure system causing widespread devastation.
Tornado	A narrow, violently rotating column of air.

6 a El Nino Southern Oscillation – the cycling of warm water between northern Australia and northern South America.
  b The Southern Oscillation Index monitors the differences in sea level atmospheric pressures between Tahiti and Darwin.
  c El Nino is a term used to describe the movement of warm surface water from the western pacific to the coast of Peru and Ecuador.
  d A negative SOI indicates low rainfall and drought.
  A positive SOI indicates high rainfall and often flooding.

7 a warmer, more, cooled, condense, particles, clouds
  b Precipitation, or rain, occurs when the droplets in clouds combine and become so heavy that air currents can no longer hold them up.
  e If clouds reach up into the higher, colder parts of the atmosphere the water droplets freeze.

8 The northern hemisphere, because the winds are rotating anticlockwise and to the right.

9 a wind
  b Temperature causes air to move vertically (hot air rises).
  Air pressure causes air to move horizontally (from a high to a low).
  c Coriolis Effect

10 a gravity b air pressure
  c cold air d cloudy, rainy weather

11 a

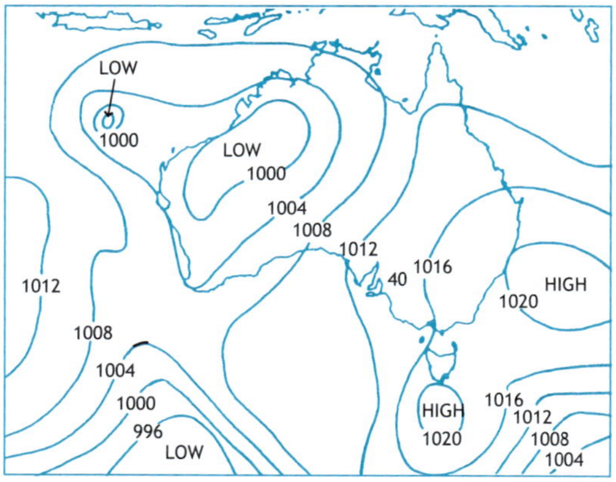

  b 996 hectopascals
  c 1016 hectopascals
  d barometer
  e isobars
  f very windy weather

12 b

Chapter 6

Science skills page 72

1 1.7 kg

2 a A b A
 c A d A

3 a The cubes will remain where it is and gravity will then pull it down into the bottle.
 b Inertia stops the sugar cubes from being swept sideways.

4

Situation	Force one	Force two
person holding	arm pulling up	gravity pulling down a briefcase
boat floating	water pushing up	weight of the boat pulling down
hot air balloon tethered to the ground	air pushing up	rope pulling down

5 a The force of gravity pulling on the block (downwards) = The force of the table pushing up (upwards)
 b Newton's Third Law.

6 a If the two students were of the same weight then they would both roll back equally.
 b For every action there is an equal and opposite reaction.

7 a & b

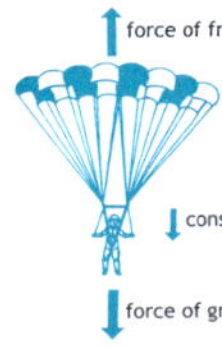

 c Because the forces acting on it are balanced.

8 The floating helium is pulling the balloon upwards but the weight of the paper clips is pulling it downwards. The balloon is not moving so these forces must be balanced.

9 c Because the car is travelling at a constant velocity.

10 a The small marble travelled the furthest.
 b Newton's Second Law of Motion.

11 Inertia of the coins keeps the column over the original spot. This is Newton's First Law of Motion. The coins fall down after each swipe due to the force of gravity acting on them.

12 a 8 newtons b 0 newtons c 5 newtons

13 a It should be a rapid outward and downward movement.
 b Things that are not moving tend to remain still unless a force starts them moving. This is called inertia. Inertia also means that things moving in a certain direction tend to keep moving in that direction

14 a 0 m/s b 50 m/s c 20 sec
 d it decreases e about 4 seconds

15 (2) There is only one force on the skydiver (weight).

C

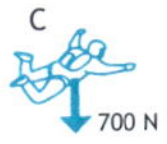

 (4) As he free falls the forces (friction and weight) become equal, so he stops accelerating and falls at a constant speed.

D

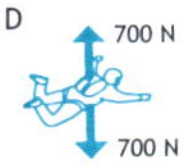

 (1) The parachute opens and the two forces (friction and weight) become equal again.

A

 (3) The resultant force (air resistance) is now upwards making him slow down.

E

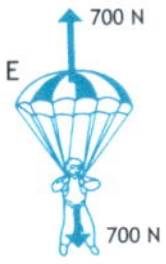

 (5) When he hits the ground the Earth pushes up and he decelerates fast.

B

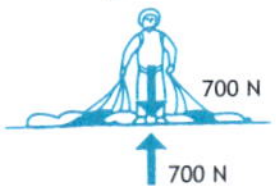

16 The hand pulls the trolley towards it but the inertia of the ball tends to keep it in the same place, so the back of the trolley comes forward and hits it. This demonstrates Newton's First Law of Motion.

17 The man is moving along at the same speed as the bus. When the bus brakes, the speed of the bus decreases. The man continues to go at the same speed, so he rolls out of the bus on his roller skates, which have reduced the friction between his shoes and the floor of the bus.

18 The force of your feet on the Earth (moves the Earth slightly backwards) = The force of the Earth on your feet (moves you forward)

19 There is a pair of equal forces acting in opposite directions — Newton's Third Law.

20 F = ma

$$a = \frac{\text{force}}{\text{mass}} = \frac{2}{10} = 0.2 \text{ m/s/s}$$

Revision questions page 76

1 forward, third, action, reaction

2 a friction b it must be constant
 c seatbelt
 d Newton's First Law — Every mass stays at rest or moves at a constant speed in a straight line unless a resultant force acts on it. The person may be forced through the windscreen of the car.

3 a The empty trolley would accelerate more quickly because it has less mass (inertia).
 b The empty one because it has less mass (inertia).
 c They would both be weightless as there is no gravity pulling on them.
 d The full trolley would still be harder to accelerate.

4 a The unit that we use to measure forces.
 b They weigh the same!

5 d

6 a You stay still but appear to slide backwards.
 b You keep moving forward even though the bus stops.
 c You are thrown to the right as your body keeps trying to go in a straight line.
 d Newton's First Law.

7 a speed, direction, movement, size, shape
 b Weight, Mass, newtons, kilograms
 c Inertia

8 a 60 kg b 10 kg c 60 kg

9 Every mass stays at rest or moves at a constant speed in a straight line unless a resultant force acts on it.

10
- Bending an object. (C)
- Balancing another force, and preventing movement. (E)
- Stopping an object that is moving. (B)
- Starting an object moving. (A)
- Changing the direction of movement. (D)
- Stretching an object. (F)

11 d

12 a The thread will break above the book as the most strain is above the book (the book's weight).
 b The thread will break below the book as the inertia of the book keeps the most strain below the book.

13 When you flick the card off the glass, the inertia keeps the coin from moving sideways. When the coin has lost its support' gravity causes it to fall into the glass.

14 For every action there is an equal and opposite reaction.

15

Term	Explanation
weight	5
acceleration	4
inertia	3
velocity	1
newton	2

16 A catapult involves the Third law of Motion.
 a The catapult moves backwards.
 b The marble moves forward.
 c The catapult hardly moves at all because of the friction between the table and the catapult.
 d The marble goes further.

17 In the absence of air friction all objects fall in a gravitational field with the same acceleration. However, air friction and weight are related. The book is hardly affected by air friction but the feather is affected greatly, so the book lands a long time before the

feather. If the feather is placed on top of the book the book leads the way through the air and they arrive together.

18 a The skateboard.
b The small child because he has a smaller mass.

Chapter 7

Science skills page 84

1 a industry b 38% for transport
2 a oil b solar
c electricity generation d transport
e solar and hydro-electricity
3

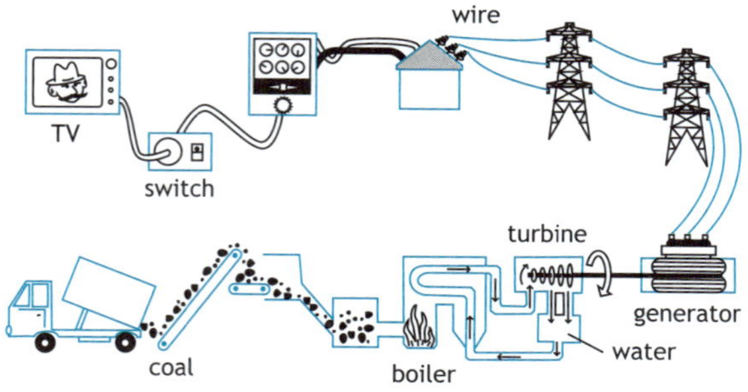

4 a electricity generation
b

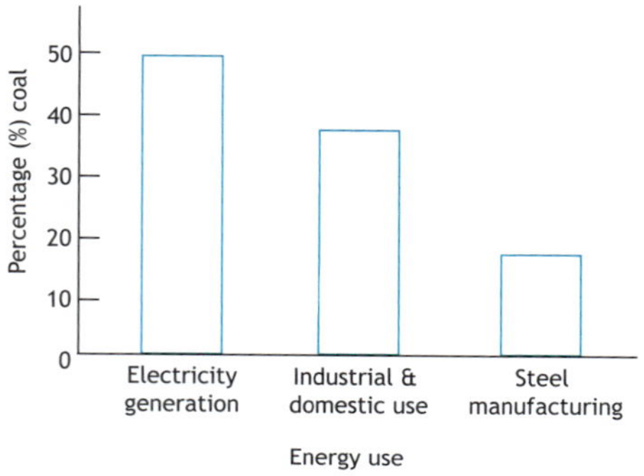

5 a transport – 30 barrels
heating – 23 barrels
manufacturing – 15 barrels
electricity – 11 barrels
fuels – 9 barrels
chemicals/rubber – 7 barrels
lubricants/paint/plastics etc – 5 barrels
b 12% c transport d fuels for ships, trains, planes etc
6 a i transport ii food consumption
b Modern societies eat more processed foods.
7 a Lights and TV on but the person is asleep. Heater on with the window open.
b Light on and the fridge and oven doors left open.
c Lights and radio on but no-one in the room.
d Heater on and tap dripping.
8
- Radioactivity from nuclear fuel. (e)
- Land degradation and other impacts such as damage to plant and animal habitats from mining, dams and waste storage. (a)
- Air pollution, both indoor and outdoor, from burning fossil fuels. (b)
- Water pollution from waste, waste storage and accidental spills. (d)
- The greenhouse effect. (c)

9 Wind energy is created as the hot air rises.

10

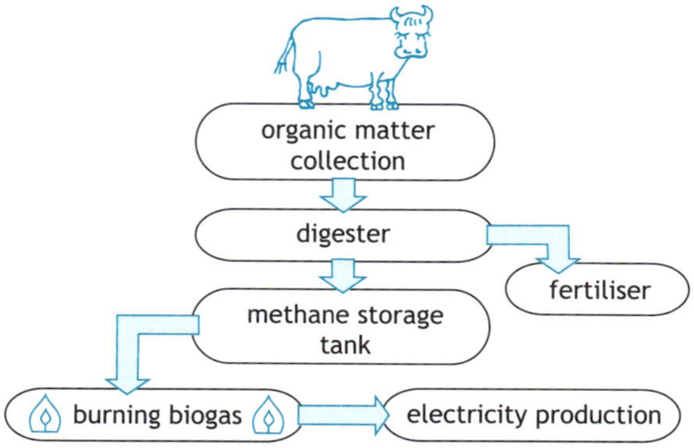

11 a Solar cells and a small motor.
b The solar cells collect the sun's energy, which drives the small motor and provides electricity to turn the wheels.
12

Source	Advantage	Disadvantage
solar	clean	needs lots of sunlight
tidal	safe	difficult technology
wind	expensive to set up	needs a windy location
hydro	has large capabilities	destroys habitats

13
- Trees to screen the southerly side from cold winds.
- Insulation to keep heat in in winter and out in summer.
- Overhang from roof to give summer shading.
- Large glass windows to absorb the heat in winter.
- Thick curtains to keep the heat in winter and keep the sun out in summer.
- Thick walls to insulate against the cold.

14 a through the roof b insulation
c use thick curtains
15 a

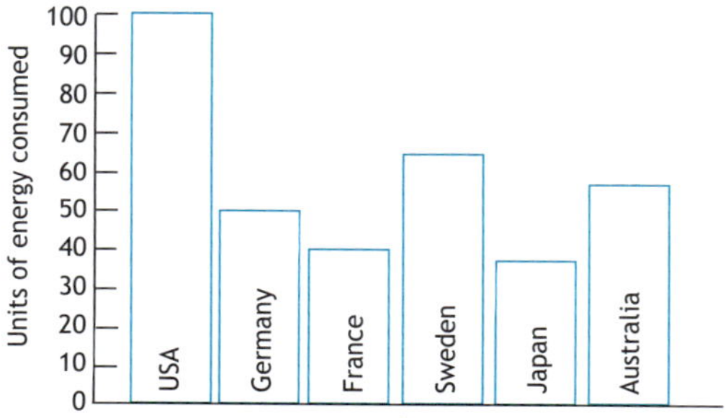

b No, they would consume a lot less energy.
16 a The amount of energy needed by the country.
b The amount of energy able to be supplied by the resources of the country.
c More coal than the country can use at that period of time. It may be exported.
d 2010
e Not enough energy for the country so it is importing oil.
f Lower the energy consumption and develop renewable energy sources.

Revision questions page 88

1 a
2 a Conserving energy means being more efficient with our energy – using it wisely, and not wasting it.
b Increase the use of liquefied petroleum gas (LPG).
Make liquid fuel from natural gas or coal.
Produce alcohol fuels from renewable plant material.
Use fuels more effectively and reduce wastage.
c Because our energy resources will eventually run out, and conserving energy now will give us more time to improve our renewable technologies.
3 c
4 Hundreds of seabirds and millions of fish died and the coastlines were polluted.
5 a carbon dioxide and sulphur dioxide
b smog
c greenhouse effect and acid rain

6 a wind power
b Yes, it is renewable.
c solar, tidal, wave, coal, etc
7 hot springs and geysers
8 water, tidal, wave, water, energy, turbines, electricity
9 a coal and hydro-electricity
b Because it is convenient and easy to access.
c i oil
ii It produces greenhouse gases when burnt.
d i wind, wave, geothermal, tidal
ii Renewable energy won't run out because it is continually being replaced.
iii There are not enough sunlight hours for the technology to be effective.
e Geothermal energy is energy obtained from the heat deep stored inside the Earth.
10 d, a, b, e, c
11 a Energy from the sun heats up the ground, which in turn heats up the air above it. Hot air rises and cold air moves in – this is wind.
b The movement of wind over the oceans creates the waves.
c The movement of air is responsible for our weather. Rain will fill the dams.
12 a nuclear and geothermal b tidal
c wave and hydro-electricity
13 a There is limited supply, and it will run out one day.
b transport c Making plastics, paint, glues etc.
14

Energy source	Is it renewable?
wind	yes
coal	
solar	yes
nuclear	
biogas	yes
natural gas	
tidal	yes
geothermal	yes
oil	
waves	yes

15 fuel, oxygen and a heat source
16 a energy efficiency b renewable energy
17 c
18 d
19 c
20 a cereal waste b household waste c sugar industry

Chapter 8

Science skills page 96

1 a 3 half-lives b 18000 years c 224 years
2 a

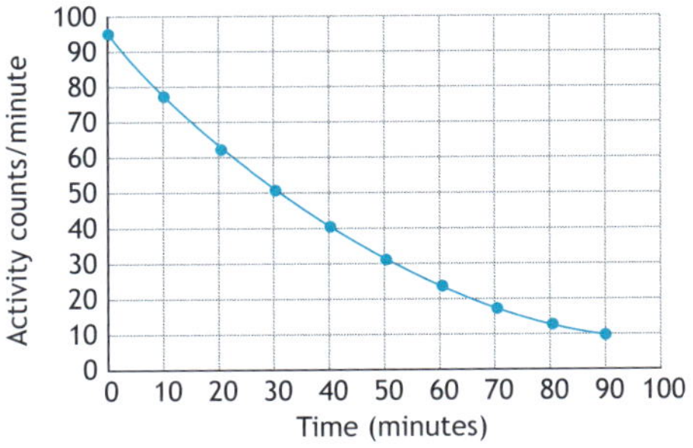

b Approximately 32 minutes – this is about the time it takes for half the atoms to decay.
3 a background radiation
b

Counts/minute	Time (minutes)
300	0
201	20
135	40
90	60
60	80
41	100
27	120

c half-life = 35 minutes

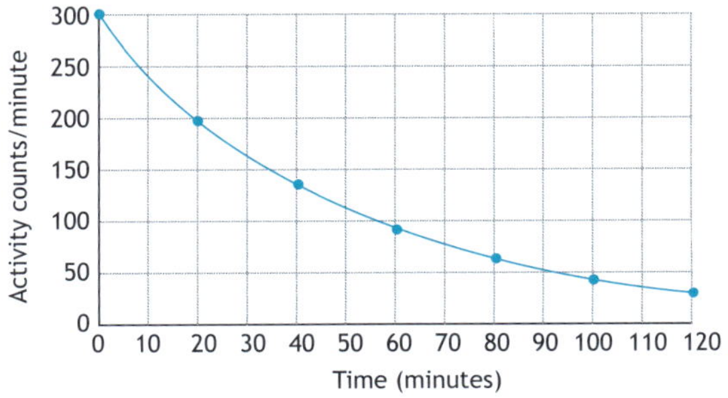

4 Nuclear reactor
↓
Nuclear energy
↓
Heat
↓
Steam
↓
Moving turbine
↓
Dynamo
↓
Electric energy
5

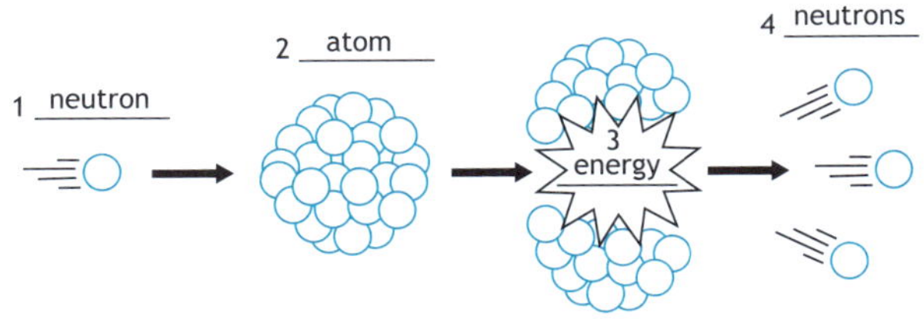

6 A – Control rods B – Fuel rods
C – Moderator D – Turbine
7 b and f
8 Fuel Uranium.
Shield Protection from radiation.
Coolant Carries heat away from reactor.
Moderator Slows down the neutrons.
Control rods Control the chain reaction.
9 a chain reaction
10 a half b a quarter c one-sixteenth
11

Properties	Beta	Gamma	Alpha
nature	electron	very short waves	positively charged
penetration	3 mm aluminium	reduced by lead	stopped by paper or skin
dangerous	yes	yes	yes

12 a 80% b 9000 million years c 1500 million years
13 a 120 b 50
14 a Lithuania b 40% c 75% d Europe
15 a beta particles b alpha particles c gamma particles
16 a iridium-192: Pipes are deep in the ground and it takes time to check the full length of the pipeline.

b technetium-99: Patient should be exposed to gamma radiation for the shortest time possible.
c polonium-210: This is not a dangerous form of radiation and it lasts for the growing time of a crop.
d phosphorous-32: This is a less dangerous form of radiation and lasts a few days.

Revision questions page 100

1 Ernest Rutherford
2 It can interfere with the normal function of cells and may lead to cancer and leukemia.
3 a Because the risk of earth movement is high and could cause a break in the shield. 100 million people will be affected by radiation.
b Should a disaster like Chernobyl occur, Australia would receive much of the escaped radiation.
4 a
5 a deuterium b no c sun and stars
d hydrogen bomb e fission reaction
f The splitting of an atom.
g atomic bomb, some submarines h yes
6 waste, thousands, rods, radiation
7 A breeder reactor uses the waste from the uranium (plutonium) as a fuel. It produces much more power and would not run out of fuel for several hundred years. Breeder reactors are, however, far more dangerous.
8 a Geiger counter b counts/minute
c It is a measurement due to natural radioactivity from the rocks of the earth.
9 radioisotope (tracer), soil, absorb, detector
10 a false b true c true d false e true
f false
11 uranium, neutrons, nuclei, neutrons, energy, chain reaction
12 d
13 a i b fusion
14 a The ability to damage the cells of all living organisms.
b It must be stored, perhaps for thousands of years, so that the radioactive particles do not escape into the atmosphere.
c It may accumulate in the food chain.
d Fusion reactors do not produce radioactive waste.
15 a from the nucleus of an atom
b alpha, beta and gamma
c half-life d yes

Test B page 102

1 a clockwise
b There is no wind from any direction.
c south east
2
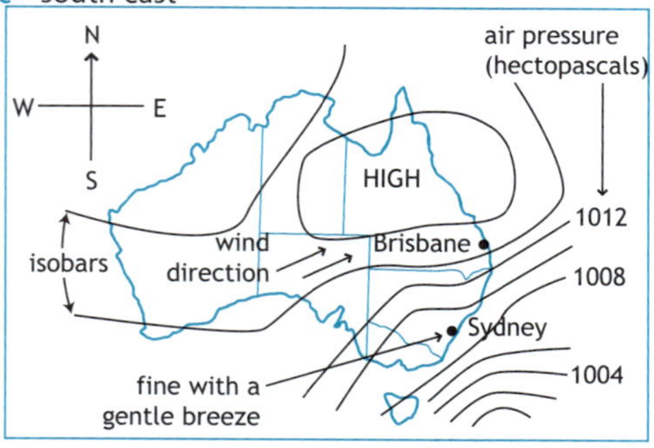

3 isobars, air pressure, hectopascals, west to east
4 a 11 °C b June c 20 °C d 25 °C
5 The maximum was 30 °C and the minimum was 3 °C.
6
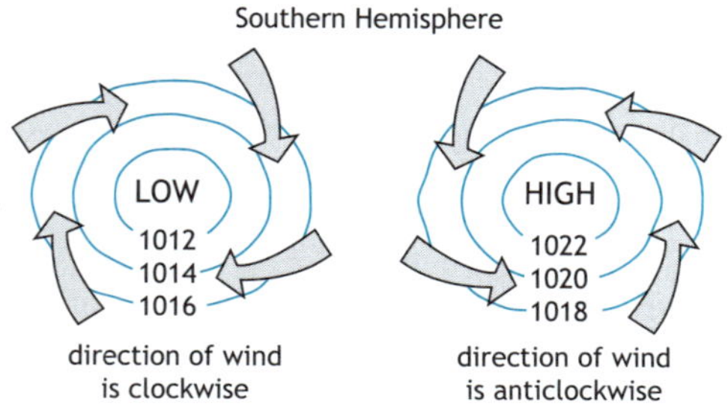

7 a Tracy = category 4
b Winifred = category 3
8 Negative phases in the SOI (drier periods) tend to be linked with reduced wheat crops and visa versa.
9 a It starts in March of the first year and ends in March of the second year.
b 6 months
c Low winter and spring rainfall causing drought conditions in summer. This is followed by late summer rains which can cause flooding.
10 wind, rises, warmed, cooled, high, low, strong, right, left, Coriolis Effect
11 a Newton's Third Law — for every action there is an equal and opposite reaction.
b To the left.
c To the right.
12 a Friction (air resistance) and driving force (pedalling).
b Because the forces are balanced.
c Unbalanced forces.
13 a 6m × 3 600 = 21 000 m/hr = 21.6 km/hr
b It will travel faster.
c It will not travel as far.
d Newton's Second Law of Motion.
14 a The steam builds up pressure and the popcorn not only busts but shoots into the air.
b Newton's Third Law
c When the hot gases in a rocket build up pressure they escape and push the rocket forwards.
15 a Newton's Third Law.
b Force, inertia, same, seatbelts.
16 The ball slides in a forward motion but the spin acts in the opposite direction and the ball rolls back.
17 a The mass slides along the table.
b The cotton breaks.
c All the books stay together and slide along the table.
d The bottom book slides out and the rest fall onto the table.
18 a
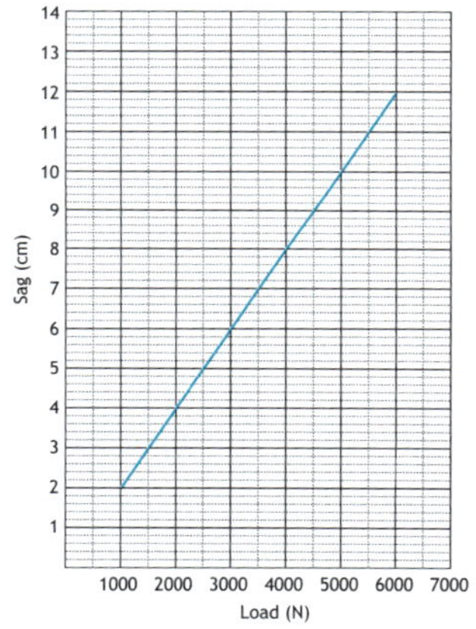

b 6.6 should be 6.4
c 9.0 cm
d 2600 N
19 a If the forces on a mass are balanced the mass will continue to move at a constant speed in a straight line.
b Clockwise.
c He is holding on tightly.
d He will have to step off and run in a straight line.
e Friction.
20 a i aluminium – B
ii paper – A
iii thick lead – C
b

Date	Mass of radioactive substance left
1 January 1990	8 kg
1 January 1995	4 kg
1 January 2005	1 kg

21 a

Time in years	Activity in counts/ minute	Activity due to the sample alone
0	124	120
1	80	76
2	52	48
3	34	30
4	23	19
5	16	12
6	12	8

b Small amounts of radiation that occur naturally and are all around.

c

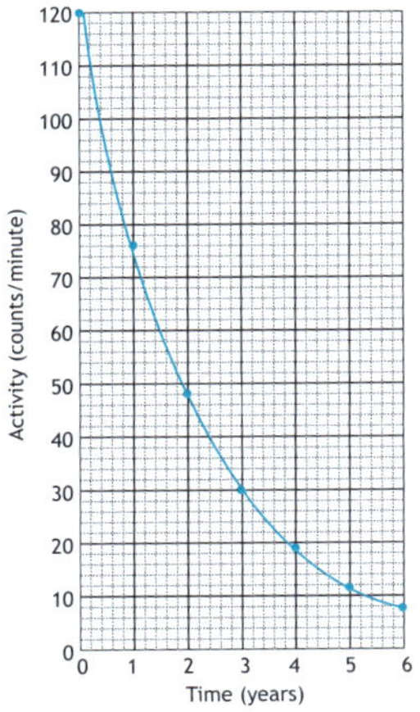

d 1.5 years

22 a Radon gas from the ground.

b medical (including X-rays)

23 The bone is $3 \times 5700 = 17\,100$ years old.

24 fossil fuel, boiler and furnace, steam, turbine, generator, electricity

25 a oil and gas

b coal

c No, we export the uranium we mine in Australia.

26

Power option	Disadvantage	Advantage
coal	pollution	easy to use
wind	expensive to set up	clean and natural
hydro-electricity	destroys habitats	reliable and efficient

27 fossil fuels, chemical, heat

28 a Dead plants covered by layers and layers of material that has squashed the plants with an enormous pressure.

b Oil and natural gas.

c Because they are made from the remains of dead plants and animals.

29 renewed / replaced, renewable, non-renewable

30 b

31 a non-renewable b renewable

c renewable d renewable

e non-renewable

32 a In summer the tree has all its leaves and so it shields the front of the house from the heat of the summer sun. In winter it loses its leaves, so the sun can shine in and warm the house.

b In summer the wide eaves create shade, and stop the sun entering the house and heating it up. In winter the sun is lower in the sky so the sun can still shine in and warm the house.

Notes

Notes

Notes